Cultural Politics of Targeted Killing

The deployment of remotely piloted air platforms (RPAs) – or drones – has become a defining feature of contemporary counter-insurgency operations. Scholarly analysis and public debate has primarily focused on two issues: the legality of targeted killing and whether the practice is effective at disrupting insurgency networks, and the intensive media and activist scrutiny of the policy processes through which targeted killing decisions have been made. While contributing to these ongoing discussions, this book aims to determine how targeted killing has become possible in contemporary counter-insurgency operations undertaken by liberal regimes.

Each chapter is oriented around a problematisation that has shaped the cultural politics of the targeted killing assemblage. Grayson argues that in order to understand how specific forms of violence become prevalent, it is important to determine how problematisations that enable them are shaped by a politico-cultural system in which culture operates in conjunction with technological, economic, governmental, and geostrategic elements. The book also show that the actors involved – what they may be attempting to achieve through the deployment of this form of violence, how they attempt to achieve it, and where they attempt to achieve it – are also shaped by culture.

Demonstrating how the current social relations prevalent in liberal societies contain the potential for targeted killing as a normal rather than extra-ordinary practice, the book will be of great use to academic specialists and graduate students in international studies, geography, sociology, cultural studies and legal studies.

Kyle Grayson is a Senior Lecturer in International Politics at Newcastle University, UK. His research interests are in the areas of political violence, security, culture, identity, and critical social theory.

Interventions
Edited by:
Jenny Edkins, Aberystwyth University and Nick Vaughan-Williams,
University of Warwick

The Series provides a forum for innovative and interdisciplinary work that engages with alternative critical, post-structural, feminist, postcolonial, psychoanalytic and cultural approaches to international relations and global politics. In our first 5 years we have published 60 volumes.

We aim to advance understanding of the key areas in which scholars working within broad critical post-structural traditions have chosen to make their interventions, and to present innovative analyses of important topics. Titles in the series engage with critical thinkers in philosophy, sociology, politics and other disciplines and provide situated historical, empirical and textual studies in international politics.

We are very happy to discuss your ideas at any stage of the project: just contact us for advice or proposal guidelines. Proposals should be submitted directly to the Series Editors:

- *Jenny Edkins (jennyedkins@hotmail.com) and*
- *Nick Vaughan-Williams (N.Vaughan-Williams@Warwick.ac.uk).*

'As Michel Foucault has famously stated, "knowledge is not made for understanding; it is made for cutting" In this spirit The Edkins – Vaughan-Williams Interventions series solicits cutting edge, critical works that challenge mainstream understandings in international relations. It is the best place to contribute post disciplinary works that think rather than merely recognize and affirm the world recycled in IR's traditional geopolitical imaginary.'
    Michael J. Shapiro, University of Hawai'i at Manoa, USA

Memory and Trauma in International Relations
Theories, Cases, and Debates
Edited by Erica Resende and Dovile Budryte

Critical Environmental Politics
Edited by Carl Death

Democracy Promotion
A Critical Introduction
Jeff Bridoux and Milja Kurki

International Intervention in a Secular Age
Re-enchanting Humanity?
Audra Mitchell

Politics and Suicide
The Philosophy of Political
Self-destruction
Nicholas Michelsen

Late Modern Palestine
The subject and representation of
the second intifada
Junka-Aikio

Negotiating Corruption
NGOs, Governance and Hybridity
in West Africa
Laura Routley

The Biopolitics of Lifestyle
Foucault, Ethics and Healthy
Choices
Christopher Mayes

**Critical Imaginations in
International Relations**
*Aoileann Ní Mhurchú and
Reiko Shindo*

**Time, Temporality and Violence in
International Relations**
(De) Fatalizing the Present, Forging
Radical Alternatives
*Edited by Anna M. Agathangelou
and Kyle Killian*

Lacan, Deleuze and World Politics
Rethinking the Ontology of the
Political Subject
Andreja Zevnik

The Politics of Evasion
A Post-Globalization Dialogue
Robert Latham

Researching War
Feminist Methods, Ethics
and Politics
Edited by Annick T. R. Wibben

**China's International Relations and
Harmonious World**
Time, Space and Multiplicity in
World Politics
Astrid Nordin

Narrative Global Politics
Theory, History and the Personal in
International Relations
*Edited by Naeem Inayatullah and
Elizabeth Dauphinee*

On the Greek Origins of Biopolitics
A Reinterpretation of the History of
Biopower
Mika Ojakangas

Insuring Life
Value, Security and Risk
Luis Lobo-Guerrero

The Global Making of Policing
Postcolonial Perspectives
*Jana Hönke and Markus-Michael
Müller*

Cultural Politics of Targeted Killing
On Drones, Counter-Insurgency,
and Violence
Kyle Grayson

Europe Anti-Power
Ressentiment and Exceptionalism
in EU Debate
Michael Loriaux

Refugees in Extended Exile
Living on the Edge
*Jennifer Hyndman and Wenona
Giles*

Security Without Weapons
Rethinking Violence, Nonviolent
Actions, and Civilian Protection
M. S. Wallace

Cultural Politics of Targeted Killing

On drones, counter-insurgency, and violence

Kyle Grayson

Routledge
Taylor & Francis Group

LONDON AND NEW YORK

First published 2016
by Routledge

2 Park Square, Milton Park, Abingdon, Oxfordshire OX14 4RN
52 Vanderbilt Avenue, New York, NY 10017

Routledge is an imprint of the Taylor & Francis Group, an informa business

First issued in paperback 2020

British Library Cataloguing in Publication Data
A catalogue record for this book is available from the British Library

Library of Congress Cataloging in Publication Data
A catalog record for this book has been requested

ISBN: 978-1-138-64605-6 (hbk)
ISBN: 978-0-367-59630-9 (pbk)

Typeset in Times New Roman
by Taylor & Francis Books

Contents

Tables

Acknowledgements

Scholarship is often a community endeavour. To these ends, I would like to thank the following friends and colleagues, who during a very long grind, expressed interest, provided feedback, and/or offered moral support: Martin Coward, Jocelyn Mawdsley, Matt Davies, Simon Philpott, Laura Routley, Una McGahern, Andrew Walton, Valentina Feklyunina, Michael Barr, Amanda Chisholm, Jesse Ovadia, Jemima Repo, James Bilsland, William Maloney, Derek Bell, James Ash, Nick Morgan, Ryerson Christie, David Mutimer, Jutta Weldes, Claudia Aradau, Jef Huysmans, Mustapha Kamal Pasha, Antoine Bousquet, Lara Montesinos-Coleman, Anna Leander, Elisa Wynne-Hughes, Andrew Neal, Tarak Barkawi, Xavier Guillaume, Ali Howell, Victoria Basham, David Campbell, Elspeth Van Veeren, Mark Salter, David Grondin, Maximilian Mayer, Colleen Bell, Rune Saugmann-Andersen, Erzsebet Strausz, Shine Choi, Saara Sarma, Juha Vuori, Klaus Dodds, Charlotte Heath-Kelly, Jeremy Crampton, J. Marshall Beier, Ian Shaw, Tina Managhan, Roland Bleiker, Elizabeth Dauphinee, Cristina Masters, Andreas Behnke, Rachel Woodward, Francisco Klauser, Heather Johnson, Ben Muller, RBJ Walker, Angharad Closs-Stephens, Pete Adey, Paulo Esteves, Ruth Blakeley, Simon Dalby, Michael Dillon, Carl Gopal, and Nick Robinson. An extra special thanks goes to Debbie Lisle, Michael J. Shapiro, and Derek Gregory who, unbeknownst to them, delivered unexpectedly kind words at particularly fraught times that kept me from abandoning the project.

The research and thinking space that enabled this book benefited from two periods of study leave in 2009 and 2012, early research assistance from Talya Leodari, and additional content analysis coding conducted by Hector Bezares Buenrostro and Paul McFadden. All of this was funded by the School of Geography, Politics, and Sociology at Newcastle University. Over the course of the project I benefited from institutional visits to Sciences Po Paris, the University of Tromsø, and the International Political Sociology Winter School at PUC-Rio. All were places where I could articulate ideas in embryonic and more polished forms. Many thanks to Hitomi Kubo, Shahrbanou Tadjbakhsh, Gunhild Hoogensen Gjorv, and Joao Nogueira for the invitations and their hospitality. In addition, I received helpful feedback from presentations and seminars delivered at Newcastle University (2009; 2010; 2012), the Standing

Group on International Relations Conference (2010), Political Studies Association Convention (2010), Popular Culture and World Politics Conference (2011; 2012; 2013; 2015), McMaster University (2011), Durham University (2013), International Studies Association Conventions (2013, 2014; 2015), British International Studies Association Conventions (2014; 2015), Cardiff University (2014), the University of Neuchatel (2015), the European International Studies Association Conference (2015), the Centre for Security Research at the University of Edinburgh (2015), and Queen's University (Belfast) (2015). Thus, I wish to thank all of those participants, discussants, panel chairs, and members of the audience who helped make this a better book. Similarly my gratitude goes to the anonymous referees of this manuscript for asking some tough questions. Any remaining errors or omissions remain my responsibility.

Special thanks to Nick Vaughan-Williams and Jenny Edkins for their enthusiasm towards the project, sound judgement along the way, and lightning fast email responses. I would also like to thank the team at Routledge, particularly Lydia de Cruz and Peter Lloyd, for answering my questions and seeing the manuscript through the production process.

Last but not least, I owe a massive debt to my family. To my mum and dad for all of their moral support over the course of this project. And to Denise, Elle, and Coco, for all of their love, understanding, and patience.

Early versions of parts of chapter one and chapter three first appeared in 'The Ambivalence of Assassination: Biopolitics, Culture, and Political Violence' published in *Security Dialogue* (2012), Vol. 43 (1), pp. 25-41. The author and the publisher wish to thank Sage Publications for permission granted under their gratis reuse licensing provisions for authors.

An earlier version of the section in chapter five on visuality and the drone debuted in 'Six Theses on Targeted Killing' published in *Politics* (2012), Vol. 32(2), pp. 120–8. The author and the publisher wish to thank John Wiley and Sons for permission granted under author reuse provisions in the original licensing form.

1 The cultural politics of the targeted killing assemblage

Introduction

On 14 February 2009 in Narsi Khel, Zangari, South Waziristan, the United States made its first attempt under the Obama administration to eliminate Baitullah Mehsud, leader of the Pakistan Taliban. While the missile attack, launched by a drone, reportedly killed 25 Uzbek militants, it did not dispatch its intended target. Moreover, the attack which was directed at a car also damaged a nearby house and a *madrassa*, killing at least eight civilians. One of the civilians killed was an eight-year-old boy named Noor Syed, who died after being hit by shrapnel that flew into his house. The United States would make another three attempts on Mehsud's life in the following months, before finally killing him in a drone strike that took place on 5 August. Over the course of these subsequent attacks, it has been estimated by human rights NGOs and investigative reporters that between 24–59 civilians were killed, including 13 children (Bureau for Investigative Journalism 2011).[1]

This series of attacks for the purposes of eliminating a 'high value' target is but an illustrative example of an emerging form of political violence practiced by liberal regimes. The use of targeted killing – the pre-meditated selection and elimination of named individuals – has become a defining feature of contemporary counter-insurgency operations led by liberal regimes in Afghanistan, Iraq, Libya, Pakistan, Palestine, Somalia, Syria, and Yemen. In particular, the use of Remotely Piloted Aircraft (RPAs) or 'drones' for the purposes of targeted killing has garnered a substantial amount of attention with questions being raised over its legitimacy and effectiveness as a security practice.[2] Supporters claim that liberal regimes such as Israel, the United Kingdom, and the United States that are involved in targeted killing operations possess the institutional capabilities to engage in a precise, proportional, and prudential manner while opponents, in light of operations like that described above, question if this is indeed the case.

While this book will contribute to these ongoing discussions, it asks a different set of questions. The aim of the analysis is to determine 'how targeted killing has become possible in contemporary counter-insurgency operations undertaken by liberal regimes'. Although its antecedents include assassination, practices of 'man-hunting', and aerial warfare, this analysis takes the starting position that targeted killing is emblematic of late-modern liberal forms of counter-insurgency and policing. That is, targeted killing can be seen as a tactic that, in part, seeks to shape social orders into preferred forms as a part of wider pacification programmes that involve both acts of violence and reconstruction programmes aimed at transforming socio-economic systems (Neocleous 2011: 197; Neocleous 2013). However, rather than expressing views on the legality/effectiveness of targeted killing, conducting process-tracing of specific decisions to target individuals, or attributing the rise in targeted killing exclusively to the development of drones, this book analyses how targeted killing is an assemblage that involves the materialisation of a set of problems across the legal, political, economic, and geostrategic domains. The objective is to map these problematisations – i.e., how people, places, things, and issues are constructed and framed as a particular type of problem as opposed to some others. These mappings then demonstrate how problem framings reflect relations of power and cultural understandings that have enabled targeted killing to emerge as a predominant practice in contemporary counter-insurgency. The central argument is that targeted killing – and the turn to drone warfare – is realised by a cultural politics constituted by common values, norms, understandings and modes of interpretation that present it as a required, if not regular, element of security provision (Hall 1997; Williams 1985). Culture not only contributes to mediation processes through which problems (past and present) are understood and addressed, but it also premediates, suggesting potential future scenarios – already in the process of becoming manifest – to which targeted killing becomes a prudential response (Grusin 2010). As such, this study proceeds on the basis that our understandings of acts of targeted killing are shaped by political, ethical, scientific, moral, and strategic ideas as well as embodied senses, and technologies, that are culturally situated. More narrowly, I am interested in schematising how political, ethical, scientific, moral, and strategic elements form a distinct targeted killing assemblage within liberal regimes.

In examining this assemblage, the original contribution of this monograph is the identification and analysis of the cultural politics of targeted killing within the contemporary geopolitical environment. The significance of the contribution is two-fold. First, it provides a substantive analysis of the conditions of possibility for targeted killing by liberal regimes that neither positions the practice as an aberration or perversion of liberal forms of governance. Rather, it shows how a targeting killing assemblage can emerge and persist within liberal regimes. Second, by taking culture seriously, it poses a series of challenges for critical strands of international relations and security studies that have become as fixated on security's technical rationalisms as those positions they seek to critique. Thus, this analysis demonstrates the value of positioning forms of political violence in relation to a broader set of discourses and material considerations.

The introductory chapter proceeds as follows. First, targeted killing is defined in relation to what is both its historical antecedent and contemporary 'other': assassination. Second, how targeted killing might fit into the general security logics of biopolitics and the liberal way of war is explored. Third, the limits of biopolitical approaches to security are discussed and a methodological approach that includes cultural analysis is forwarded. Fourth, building upon this methodological approach, assemblages are introduced as a way of capturing the inter-relationality of discourses, institutions, materiality, and power-relations contributing to targeted killing. Fifth, the importance of problematisations within the targeted killing assemblage is presented. Finally, the chapter ends by revealing the lines of flight for the substantive chapters to follow.

Defining assassination and targeted killing

What are understood as assassination and targeted killing in security discourses and what are their relationships to the practices of contemporary liberal governance? In seeking answers to these questions, scholarly analysis and public debate has primarily focused on four issue areas. The first has been the legality and ethics of assassination and targeted killing. This has been broken into two subfields in the literature: those that examine the relationship between these practices and international law (e.g., Enemark 2013; Finkelstein et al. 2012; Rae 2014; Strawser 2013; Gregory 2015) and those who evaluate the democratic legitimacy of related decision-making structures within regimes like the United States and Israel (e.g., McCrisken 2013; Falk 2014). The second has been an interest in how these forms of violence are undertaken including the tactics used, the security institutions involved and their relation to broader notions of grand strategy and counter-insurgency (e.g., Williams 2013; Kaag and Kreps 2014; Rogers and Hill 2014; Sloggett 2014). The third has been their effectiveness and consequences including the direct impacts of assassination and targeted killing on insurgency movements and broader geopolitical environments (e.g., Hafez and Hatfield 2006; Iqbal and Zorn 2006; Kober 2007; Zussman and Zussman 2006; Johnston 2012; Wilner 2010; Jordan 2009; 2014; Hepworth 2014; Freeman 2014). The final focal point are critiques that have emerged from these three areas whose commitments run from defences of due process, to concerns over the liberal sovereign produced by individualised killing, to pragmatic claims that any immediate potential effectiveness is trumped by longer-term geopolitical costs, to the dangers of 'cost-free war' where violence is administered from a distance (e.g., Carvin and Williams 2015; Chamayou 2015).

Although it will be shown below that the supposed exigencies of global counter-insurgency have renewed discussions of assassination and targeted killing in liberal political contexts, it bears noting that they have never been completely absent in legal, political, and strategic discussions. More broadly, despite continuing uncertainty over its ultimate effects, assassination – as a precursor to targeted killing – played a pivotal role in global political history and the representational practices that shape the way in which global political history is understood. Assassination has served as a marker for the boundaries of legitimate ethico-political conduct and military practice. While celebrated in ancient Greece and Rome as the highest possible form of public service in the defence against tyranny, assassination has often held less prestigious positions within Judeo-Christian moral codes (e.g., see Ford 1987). However, condemnation has co-habited with the strategic realisation that assassination may have potential as a political and/or security tactic directed towards the elimination of threats, rivals, oppressors, or tyrants. As such, within the broader Western philosophical and legal tradition, assassination – often presented as 'regicide' or 'tyrannicide' – has been deliberated upon extensively by theorists such as Saint Augustine, Thomas Aquinas, Niccolo Machiavelli, Hugo Grotius, and Thomas Hobbes in attempts to understand its socio-political dynamics, probable effects, and normative composition.[3]

Commonly, assassination has been used to denote the 'murder of a political elite by an individual who performs the act in a non-governmental role' (Kirkham et al. 1970: Appendix A quoted in Ben-Yehuda 1990: 347–348). Many also consider the extra-judicial killing of an individual for political motives by the state apparatus or its agents to be a form of assassination (Stein 2003). Others wish to delineate between assassinations of state and non-state actors. Nachman Ben-Yehuda (1990: 348–349) notes that the latter – political assassinations sponsored by the state – should be more precisely labelled as 'political executions'. But as Chris Downes (2004: 152) argues, assassination has also conventionally implied that the killing has transpired 'through treacherous or perfidious means' whether it has occurred during times of conflict or relative peace (see also Guiora 2004; Bazan 2002; Kretzmer 2005). As a result of this association, assassination has often been coded as a tactic of the weak, and at times, has been overtly gendered – like poisoning – as a feminine practice outside of the norms of the legitimate spectrum of violence (Thomas 2000; Price 1997; Stocker 1998; Hallissy 1987).

Thus assassination, as will be shown in chapter 3, has had an ambivalent position within security discourses. Some states that have openly engaged in assassination are trying to distance themselves from its negative associations with respect to treachery, perfidy, and femininity. Borrowing from medical discourses on cancer and HIV/AIDS treatments, their preferred terms are 'targeted killing' or 'targeted selection'. These are used to describe the intentional selection, pursuit and eradication of specific individuals in security operations, through special forces operations, sniping, car bombing, or forms of aerial attack, including drone strikes (O'Brien 2001: 108; Kober 2007). By intertexting with medical discourse and potentially circumventing the less savoury connotations of assassination, targeting killing implies a level of selectivity, precision, and technical mastery in initiating and performing the decision to exterminate. The implication is that these targeted individuals are themselves akin to cancers who must be terminated before they proliferate and further sicken the body politic. As 'intractable individuals unable to govern themselves according to the civilized norms of a liberal society of freedom' these people become subject to permanent incapacitation (Rose 2007: 248–249). And the associated claim is that those states who label their assassinations or political executions as 'targeted killings' possess the technical and methodological acumen to obtain reliable intelligence, make decisions, and act upon them in a way that minimises indiscriminate harms as outlined in humanitarian law while lowering their levels of vulnerability to a debilitating attack (see Schwarz 2015).[4] Thus, Kessler and Werner (2008: 290) argue, 'targeted killing fit[s] the logic of precaution and risk management that dictates contemporary security policies' by serving as 'a specific form of "uncertainty absorption"… aimed at eliminating dangers.' In contrast, assassination with its linkages to treachery and femininity implies motivations, tactics, and goals that are charged, emotional, and imprudent. As such, assassination functions as targeted killing's other in security discourses, a claim that will be substantiated in chapters 2 and 3 to follow.

The very act of labelling an act of violence as an assassination or targeted killing is a choice whose rationale will be shaped by predominant discourses, modes of understanding, and socio-political contexts. They are essentially contested concepts. Therefore, Ben-Yehuda (1990: 348) argues that assassinations are best seen as a rhetorical device for the purposes of 'socially construct[ing] and interpret[ing] attempts to kill a specific social actor for political reasons [that] hav[e] something to do with the political position or role of the victim [all of which can be connected to symbolic moral universes].' By labelling a killing as an assassination – or targeted killing – the act, actors, and associated effects are imbued with a level of importance and status that is often absent in other forms of political violence or extermination. However, while the assumed symbolic importance of the victim may be similar, targeted killing is now presented as the polite face of named termination, one that seeks to avoid assassination's treacherous, perfidious, and potentially feminine connotations.

While the myriad ways that assassination and targeted killing are defined, branded, and/or constituted as objects of study are important, these are them- selves located within a broader matrix of security understandings, practices, and horizons. Thus, rather than fixate on defining what assassination and targeted killing is – or ought to be – it is important to focus on what they do and how they have become incorporated as specific types of problems within con- temporary security discourses. More precisely, what might the appearances of assassination and targeted killing tell us about how security and security threats are being identified, measured, assessed, and managed by liberal regimes engaged on the frontlines of global counter-insurgency? And what might assassination and targeted killing tell us about how liberal regimes perceive themselves?

Biopolitics, security, and the liberal way of war

The questions posed above direct inquiry away from orthodox understandings of targeted killing as a response to outside stimuli and instead conceive of the practice as a framework that makes these stimuli perceivable to us. As Miller and Rose (2008: 14–15) argue this suggests that:

problems [to be addressed through targeted killing] are not pre-given, lying there waiting to be revealed. They have to be constructed and made visible, and this construction of a field of problems is a complex and often slow process. Issues and concerns have to be made to appear problematic, often in different ways, in different sites, and by different agents...[And] the activity of problematizing is intrinsically linked to devising ways to seek to remedy it.

How then is the targeted killing assemblage related to liberalism? And how do problematisations of targeted killing reflect the broader cultural politics of regimes that claim to adhere to liberal forms of governance?

Mitchell Dean (1999) has argued that it is difficult to reduce liberalism to a finite set of ideas or institutional structures. While we may most closely identify liberalism with things like the 'rule of law', individual rights, negative liberties, or democracy, if relations of power become our primary reference point, it makes sense to view liberalism as a particular way of posing problems or as a problematising activity (Dean 1999: 49). Historically, we can point to liberalism's critiques and problematisations of absolute monarchy, of the worst excesses of laissez-faire capitalism in the late nineteenth century, of socialism, anarchism, and fascism, the welfare state, the nation state, and even more recently several brands of fundamentalisms.

Therefore, on the one hand, liberalism has shown itself to be concerned with the possibility of too much government. It assumes that there is a non-political sphere (e.g., the economy, civil society, family relations, or health) that is governed by laws of nature and processes that are – and should be – completely separate from sovereign authority. At the same time, it recognises that the well-being and proper functioning of this sphere is absolutely necessary for effective governance. Such a conceptualisation creates a tension. Liberalism then as a form of rule takes both freedom and the 'natural' aspects of life as its essential reference points for governing (Duffield 2007: 4–8). This often requires interventions – usually framed not as matters of politics but of technical and/or scientific management – into the most routine aspects of everyday life. Although these interventions may confer benefits, they also involve relations of power; these are *biopolitical* interventions that seek primarily to administer the processes of life at the aggregate level of population (Duffield 2007: 6). Governing through freedom requires that behaviour, its impacts, and effects, be monitored, determined, and acted upon when necessary. The argument is that freedom itself – as a necessary requirement of liberalism – depends upon forms of biopolitical management to ensure that it does not create conditions – and subjects – that are not amenable to being governed. Governing through freedom thus requires that behaviour, its impacts, and effects, be monitored, determined, and acted upon when necessary. Above all, ungovernable fields must not be cultivated. Thus, in certain instances, prudential interventions are required into domains of life that might otherwise be allowed to run their natural course (de Larrinaga and Doucet 2008).

The logic that contributes to the possibility of targeted killing is therefore reflective of longer standing changes to the ways in which mechanisms for governing have been viewed by those who govern. Previously, I have noted that Michel Foucault outlined in a series of lectures at the end of the 1970s, that what it means to govern begins to transform in the eighteenth century in western Europe as the means and ends of government shifted from being primarily concerned with 'the (geopolitical) aggrandizement of the state towards the mastering of population as an instrument through the use of newly developing statistical sciences' (Foucault 2003: 244; Foucault 2007 quoted in Grayson 2008: 385).

He argues that in this context, one sees the partial transposition of the Christian pastoral ethos into the development of procedures of statistical knowledge creation to deal with problems such as famine. If one reads *Security, Territory, and Population*, Foucault (2007) traces these developments, showing all the starts, stops, and contingent adaptations that occurred along the way. Identifying advantages in populations that had to that point remained hidden and miti- gating those underlying risks and/or uncertainties that could generate disruptive manifestations became both the definition of, and goal for, effective governing. What Foucault referred to as 'governmentality therefore required that a whole novel series of security apparatuses and complexes of knowledge – particularly the statistical sciences – be utilised to improve the welfare, living conditions, health, wealth, and longevity of population' (Grayson 2008: 385).

While these developments in public welfare were often related to 'humanitarian' or socio-moral intuitions about what constituted legitimate government, they also had a martial face, and were related to new initiatives such as universal con- scription as well as socio-economic developments like increased urbanisation and industrialisation. With the survival of the state and existing social order in the European context tied to notions of territorial defence as well as maintaining order internally, the number of men who had to be excused from military service as unfit, how to best control the growing *lumpen proletariat*, and how to organise society to maximise its labour potential were contributing factors in the establishment of these programmes. As health, wealth, and the well-being of population became a priority for government, public sanitation, universal education, the growth of psychiatry and criminology, the professiona- lisation of key occupations such as medicine and policing, the development of public insurance schemes, and the problems that these measures were said to be solving became central concerns for government. This 'biopolitics…a means to rationalize the problems presented to governmental practice by the phenomena characteristics of living human beings constituted as a population…was also an ends that practices of governmentality sought to achieve, that is to provide a compelling rationalization of the phenomena characteristics of populations' (Rabinow and Rose 2003: xxix quoted in Grayson 2008: 385).

Biopolitics identified specific phenomena that could and should be admi- nistered by the state managerial apparatus (e.g., communicable disease) while constructing through its truth-telling practices specific classification schemes that identified sub-sets of the general population that needed to be directly managed: 'the aim was to reduce the prevalence and potential impacts of what were represented as their associated phenomena' (Grayson 2008: 385). Deviations from the norm of acceptable ways of life, their prevention – or correction if efforts failed to prevent their manifestation – became the central problematique for governing. The art of government therefore required prudential balancing to promote the 'right ways' of living in 'a system anxious to have the respect of legal subjects and to ensure the free enterprise of individuals' (Foucault 2003: 202 quoted in Grayson 2008: 385).

It has been contending accounts of what constitutes proper biopolitical management that have helped to define differences between liberalism and other forms of governance as well as differences within forms of liberalism itself. Despite shifts amongst various forms of rule during the twentieth century – liberalisms, fascisms, socialisms, welfarisms – all shared a presupposition that 'the real [world] is programmable by authorities' (Miller and Rose 2008: 211). All believed that 'the objects of government' were constituted such that problems and shortcomings would be 'amenable to diagnosis, prescription, and cure' (Miller and Rose 2008: 211). Differences amongst these systems could be found in the extent to which intervention was required (including in which domains) and what kind of population was to serve as the referent object to promote particular ways of living. Nevertheless, various forms of rule developed a 'will to govern', a desire to invent new strategies of governing that could succeed where others had failed (Miller and Rose 2008: 211). One of these strategies has been security.

Those who have conducted genealogies of security and warfare, such as David Campbell (1998), James Der Derian (1995), Michael Dillon and Julian Reid (2009), and Mark Neocleous (2008) have suggested security and defence as elements of governing both constituted and reflected dominant corpuses – scientific, national, diplomatic, legal, economic, and cultural – at a given time. Security policy, which Foucault argued developed as a form of biopolitics, begins to be understood primarily 'as a series of political rationalities and technologies with the aim of regulating circulation in order to manage contingency' (de Larrinaga and Doucet 2008: 524). Over time, the advent of new technologies and forms of community identity within the restrictions offered by internationally recognised juridical sovereignty combined with concepts from the life sciences – the problems of adaptation and mutation featured prominently within these security logics. From these logics, targeted killing can be understood, in part, not as a judicial sanction, but as a potential recourse for controlling movement away from what are recognised as liberal ways of life, that is, as a way of eliminating emerging ways of being that might threaten the predominance of liberalism.

The emphasis on adaptation and the fear of dangerous mutation is argued to have resulted in a dramatic increase in the perceived magnitude of potential risks at a time when the ability to mitigate them through the traditional practices of national defence was diminishing (Dillon and Reid 2009). The myth of national invulnerability through security provision faded in the aftermath of total war and was supplemented by nuclear weapons development and the establishment of a Cold War security architecture rooted in a balance of terror through mutually assured destruction. Without hyperbole, the consequences of war could now threaten humanity in its entirety. In contrast, Nikolas Rose (2007) has shown that concurrently in the human sciences, discovery and advancements in knowledge of the human genome had ramifications that began to shift into everyday socio-political life. Increasingly – albeit sometimes in haphazard forms as witnessed by the recent financial crisis – the role of government and the very definition of personal responsibility became framed in terms of the problems stemming from susceptibility and the search for enhancement. In other words, the key to good conduct of governments, populations, and individuals is to determine their susceptibilities and then partake in activities, treatments, or policies that had been determined to enhance characteristics, skills, technologies, and/or social structures that will prevent susceptibilities from transforming into full blown threats with negative social and/or individual consequences. Thus, the proliferation of CCTV cameras, anti-bacterial cleaning agents, self-checking credit services, medications for the treatment of latent conditions, and the new requirements to be satisfied for citizenship, are developments that can be associated with this gradual shift in how security is provided. And it is this change in attitude that has constituted the contemporary ethics of liberal humanism: the ethos of enhancing capability and resiliency through the suppression of susceptibility.

Michael Dillon and Julian Reid (2009) have argued that this conceptualisation of security, one motivated both by liberal humanism and biopolitical desires to enhance individual capabilities within the right-kind of populations, has led to the emergence of what they call the 'liberal way of war'. While the rhetoric may focus on the spread of democracy, free markets, and human rights, the referent object to be secured in this way of war is a biological being defined in terms of its characteristics and its capacity to be a productive contributor to the health of the species. For Dillon and Reid the 'species' are the individuals – and practices or ways of life – who, in aggregate, form the sum of the population subject to liberal forms of governance (Dillon and Reid 2009: 18–33). They argue the constitution of the species has increasingly been determined through 'information as code' – compiled through a range of knowledge production practices from the life sciences to religious studies – rather than by function (Dillon and Reid 2009: 22). Classification thus proceeds from range of information about you as a biopolitical subject, not necessarily by how you have behaved thus far.

The information considered pertinent to your subjectivity can be biological information (e.g., how predisposed are you to weight gain?), sociological information (e.g., what is your class background), psychological information, even geographical information. This information is not primarily analysed at the level of the individual. Rather it is used in aggregate to compile 'profiles' that in turn can be understood as 'sub-populations'. These include not just the characteristics that make up the profile but also risk calculations and behavioural propensities said to define the sub-population. It can be harnessed for the purposes of identifying 'high value' targets, those whose perceived capabilities are understood to make them particularly dangerous. For contemporary forms of liberalism, a key to effective governance is to always be collecting this information, making calculations, and monitoring data for trends that indicate changes to norms or the formation of abnormalities, deviances, or unproductive adaptations that threaten the overall health of the species. When a living thing is defined in terms of information as code, it implies that this code can be cracked, making bespoke recoding and recombination possible for new purposes or even to maintain old ones. With recent advances in the life sciences, there is now a general consensus that the normal status of individual organisms and the systems they belong to is adaptively dynamic, constantly in flux, changing, adapting, relearning, and reapplying. Thus, Dillon and Reid (2009) argue that what we see in liberal regimes – both with inward and outward facing forms of governance – is a constant auditing of life to determine what forms – as well as their associated traits and features – pose risks to species life, as understood as a product of liberal forms of rule. The computation of these formulas and what these sums are understood to represent open the possibility for the targeting and elimination of individuals (or populations) that have deviated too far away from the acceptable bandwidth of tolerable risk.

These particular understandings of 'life', of individuals, of populations, and of phenomena produce a discourse of danger that centres on traits and features which are said to indicate a potential danger at some future time as opposed to specific actions that directly threaten in the here and now. Intervention is not limited to the manifest; it operates at the very level of possibility. These include the possibility that natural adaptations and change can go wrong; the possibility that there may be resistance to productive changes brought forward through forms of intervention – or that a population is incapable of 'learning' (what is best for them); and the possibility that a localised abnormality or resistance proliferates and disrupts the entire system. Dillon and Reid (2009) argue that the governing logic is suspicious of anything that approaches what have been determined to be limits of normality. From a biopolitical standpoint, activities outside of pre-established norms become seen as threats in part because by falling outside of these norms, it becomes increasingly difficult to discern probabilities and therefore control any potential risks that may be generated from them. Mark Duffield (2007: 19–24) refers to this as the distinction between insured (measurable/knowable) and uninsured (unmeasurable/unknowable) life and argues that these two designations represent clear delineations in the global governance of populations with considerable political effects. The former may need to be protected while the latter must be transformed (into the former), eliminated, or allowed to dissolve through other mechanisms which require no active input. The liberal way of war therefore becomes a form of policing at both the level of population and the level of the individual.

Within the biopolitics of the liberal way of war, Dillon and Reid (2009: 26) argue that the metric of good rule is purported effectiveness not legitimacy – as defined by the ideals of orthodox liberal political theory. But, there still remains a firm conviction in the superiority of liberal values and technologies of governance. When actions appear to contradict these values, it is argued that they are presented as unfortunately necessary, temporary, exceptional, and proportionate in relation to the scope of the threat faced. Thus, as they note, the liberal way of war operates through complex forms of truth-telling that extend beyond the life sciences, including legal reasoning. Through a diffuse set of practices, the liberal way of war produces an interesting governmentality. Within contemporary geopolitics, part of the importance of the liberal way of war comes from how it has normalised pre-emptive war or what is also called anticipatory self-defence (Massumi 2007). In doing so, it has forwarded a set of processes that define and execute necessary killing of deviant ways of life. This can involve active events like targeted killing; in other instances it involves passive practices of letting die (i.e., not saving/preserving life). In doing so, Dillon and Reid (2009: 6–7, 104) argue that the liberal way of war creates an unresolved paradox about what governing needs to be prepared to do: to wage war not just in the name of the humanity but on this very humanity if necessary. But are the practices contributing to this paradox exclusively produced by the biopolitical logics of the liberal way of war? What is left out by this account?

Beyond liberal biopolitics: a return to culture

In addition to Dillon and Reid, many analysts have presented compelling arguments about how the liberal credo of governing contingency through freedom can, at key junctures, transform into a politics of the exception whereby the most basic freedoms and protections are stripped away in the face of what is defined as an existential threat (e.g., Salter 2006; Neal 2006; Edkins et al. 2004; Aradau and Van Munster 2009; Dauphinee and Masters 2007; Closs Stephens and Vaughan-Williams 2008). Whether through the spectre of 'bare life' or 'information as code', these analyses of contemporary security governance and the practices of liberalism potentially offer a rich terrain for an account of the (re)turn to targeted killing and its political consequences (Agamben 1998; Dillon and Reid 2009). Although agreeing in principle with these assessments of what becomes possible within liberalism, my argument is that while biopolitical accounts provide important – and competing – insights into the processes that turn targeted killing into a mechanism of security provision, it is also critical to account for the politico-cultural significance of targeted killing beyond its positioning as an 'exceptional' measure within security discourses.

There are four reasons for this view. This first, as will be shown in the analysis that follows, is that the exception as a normative suspension of the law at best explains very little about what makes targeted killing possible in liberal regimes and, at worst, ignores the broader relations of power relevant to understanding this form of political violence. The second is counter-intuitive: the focus on exceptional security measures engendered by biopolitics is depoliticising. The biopolitics–exception nexus directs focus onto either the intricacies of the law and/or the scientific thought underpinning contemporary security practices without considering the socio-political contexts from which they emerge and within which they operate. The third is that these approaches fail to capture the breadth of the geopolitics underpinning contemporary security governance. This is achieved by overlooking how quotidian and exceptional places interact – the emphasis is always on the latter. Overlooking the quotidian is important because it results in compressing the terrain where subject positions are produced through biopolitics in general, and targeted killing in particular. Finally, even in critique, too much credit is often granted to the capability of liberal regimes to police and control populations through security governance. Thus, myriad ways that contingencies shape the commissioning of political violence are overlooked. And one of the sources of contingency is culture.

Methodology and culture

The above discussion of how targeted killing has become possible is only preliminary and will be explored in further detail throughout the monograph. What will be shown in the analysis that follows is that within certain liberal contexts there has been a resurgence in targeting killing as either a discursive nodal point or security performance. For example, in 2014 researchers for the Council on Foreign Relations estimated that the United States had been involved in its 500th targeted killing operation by drone, a conservative calculation of the scope of the practice given this did not include special operation raids (known as kill/capture missions) or any operations conducted inside Afghanistan and Iraq (see Zenko 2014 and Masters 2013). The resurgence in targeted killing can, in part, be linked to transformations in state and societal tolerances of perceptions of vulnerability in the security sector. The growing intolerance must be understood initially within the context of the global 'war on terror'; however, while 9/11 certainly accelerated these processes, it would be overly simplistic to identify it as the primary cause. The transformations themselves are a product of a complex set of factors, including the incorporation of broader biomedical logics centred on notions of susceptibility and enhancement; a geographical imagination oriented towards the exposure of clandestine networks; fears of risk proliferation across borders; gender and sexual norms; and the continuing influence of Orientalist understandings of the Muslim other which are subsumed within these logics, imaginations, and fears (Tuastad 2003; Puar 2007). At the same time, those accounts that stress risk and biopolitical elements of contemporary liberalism provide an insufficient understanding of how risk and biopolitics are culturally embedded practices that necessarily draw from wider representations for their performative force.

As such, this monograph provides a cultural analysis of targeted killing. By cultural analysis, I am referring to a mode of enquiry that seeks to understand how political, social, legal, and economic outcomes are constituted and mediated through culture. I understand culture to include systems of signification, material objects, ways of feeling, sensing, and being, ideas, values, systems of belief, regimes of truth, and modes of interpretation that are understood to be held in common by a collectivity whose bonds are both dense and imaginary (Hall 1997; Williams 1985; Weldes et al. 1999). Culture is also not a thing that operates in isolation or as a black box; it is relational and how it is understood can often be as much in terms of difference in relation to its others – cultural and otherwise. At the same time, I do not wish to elevate culture to the status of a prime-mover or Archimedean circuit comprised of static elements through which all political activity is generated and made meaningful. Rather, it is the plasticity of culture that is of interest. More specifically, it is the paradox of culture, the contest between malleability and fixed forms that is important to understanding the re-emergence of targeted killing as a type of political violence. On the one hand, culture is constantly shifting, driven by a dynamism generated by positive feedback loops that produce changes at the margins which can then lead to systemic level transformation. On the other hand, this dynamism is juxtaposed with inclinations towards negative feedback loops of constructed traditions, erasures, and effacements that are central to its reproduction. How targeted killing has become possible, but also institutionalised, at the start of the twenty-first century requires that emerging security thinking based on advances in science and the entrenchment of a historically specific form of governmentality be located within longer-standing notions of what is permissible to do to others in the name of honour, religious belief, and the impulse to civilise. What one may be permitted to do and to whom this is to be permitted are questions mediated through culturally shaped logics, techniques, narratives, understandings, and forms of embodied sensibility that underpin these directives. By setting culture aside, accounts that stress risk and biopolitical elements of contemporary liberalism on their own terms provide an insufficient understanding of how problematisations of risk and biopolitics are culturally embedded practices that necessarily draw from wider representations for their performative force and structural violence. The argument therefore is that to understand how specific forms of violence become prevalent, it is important to determine how problematisations that enable them are shaped by a politico-cultural system in which culture operates in conjunction with technological, governmental, and geostrategic elements.

The analytic emphasis on the development of problematisations in relation to culture is one based on claims that particular problematisations arise given the presence of certain cultural factors which in turn may then be reshaped themselves through processes of management and/or resolution. It is not a claim to an innate determinism (i.e., particular cultural factors *invariably* produce specific problematisations) or a reification of culture as an independent variable of last resort to explain otherwise unexplainable behaviour. Rather, the analysis infers that targeted killing as form of security management increases in *probability* given a set of factors (see Protevi 2009; 2010). The analysis seeks to demonstrate how the specific problematisations that make targeted killing possible and shape its governmental regulation necessarily go beyond the more commonly identified rationalities underpinning counter-insurgency. It also wishes to show that the actors involved, what they may be attempting to achieve through the deployment of this form of violence, how they attempt to achieve it, and where they attempt to achieve it are also shaped by culture. Thus, where Protevi (2009: location 405) deploys develop-ment systems theory, with its emphasis on 'on the life cycle [of the organism], developmental plasticity and environmental co-constitution', as a means to explore assemblages of politics, affects, and the constitution of subjectivity, I wish to focus on the specific form of violence (targeted killing), the plasticity of its constitutive elements, and its politico-cultural co-constitution. The point is to demonstrate how the current social relations prevalent in liberal societies contain the potential for targeted killing as a normal rather than extraordinary practice.

This study is not concerned with tracing the policy-process in key institutions to provide an evidentiary chain of linear causality that explains why targeted killing takes place given a set of variables. Nor does it seek to reduce the complexity of political decisions and cultural proclivities by nominating specific individuals as the primary agents of the shift to targeted killing within counter-insurgency policy. Claims to linear cause and effect explanations provide narratives that must either reduce the complexity of cultural practices to factors that explain very little much of the time or become embedded within the discursive formation that is the object of analysis in order to explain a lot some of the time. In particular, while the former's reductionism may be quantitatively elegant but politically unenlightening, the latter approach risks becoming trapped into modes of understanding held by those in positions of authority. Not only does this jeopardise the ability to critically engage with these modes of understanding and the institutions through which they are distributed, it also increases the probability of becoming enthralled with them, buying into their logics, their mystifications, and their myriad modes of violence.

As such, the concern here is to answer the question: how has targeted killing become possible? What discursive formations, styles of thought, culturally located understandings, and technological assemblages provide the potential for the killing of named individuals by liberal states in counter-insurgency? While particular individuals have played a central role in this process, they have been travelling across a landscape that is more vast and varied than their internal views, opinions, motivations, and ambitions. It is problematisations in politics, law, economics, and military strategy that are culturally mediated such that they enable configurations of power which give license to practices like targeted killing. This book seeks to uncover these problematisations, the regimes of truth that underpin them, and the forms of political subjectivity generated through the violence of targeted killing.

In doing so, a variant of the concept of assemblage is put forward to guide the analysis (see Deleuze and Guattari 1987; DeLanda 2006). As a general organising concept, Eric Sheppard (2008: 2607) has argued that an assemblage is 'a whole whose properties emerge through interactions among [heterogeneous] components'. As such, assemblages are relational and multi-scalar, yet they are not over-determined by their relational properties; they contain emergent properties that are not a product of underlying essentialisms. They 'are socio-natural, with agency operating in all domains [as] components play roles that vary from material to expressive in nature' (Sheppard 2008: 2607). Changes within assemblages are fostered through processes of repetition that can lead to minute differences in configuration through which new relations emerge. As Colin McFarlane (2011: 383–384) notes, 'assemblages set the machine running. It is in this immanent causality that power produces new composites and effects, including discordant parts that are lived as assemblages'. Thus the components of an assemblage shape its form through mechanisms that foster homogeneity (i.e., territorialisation) or that undermine it (i.e., deterritorialisation). Their relations are contingent and Sheppard (2008: 2607) notes, are best revealed through empirical investigation.

The point of using assemblage as a conceptual framework is to 'account for a *synthesis* of the properties of a whole not reducible to its parts' (DeLanda 2006: 4). The emphasis is therefore on 'the actual exercise of capabilities' through *relations of exteriority* whereby capacities exceed the properties of individual components because individual component capacities 'involve reference to the properties of other interacting entities' (DeLanda 2006: 10–11). Thus linkages between components within an assemblage are not '*logically necessary*' but rather '*contingently obligatory*' and the components themselves may be removed and placed into a different assemblage where they may contribute to different dynamics (DeLanda 2006: 10–11).

Assemblages can be conceptualised as involving three axes: one that defines the role played by parts as purely expressive (on one pole) and purely material (at the other pole), recognising that a component may play multiple roles by making use of different capabilities. Expressive modes are to be understood as encompassing more than formal modes of communication. It is along this axis that we see technologies such as drones, economic structures, counter-insurgency discourses, and communicative forms that express political violence like kinetic strikes.

The second dimension according to DeLanda (2006: 12) is defined by the:

> variable processes in which these components become involved and that either stabilize the identity of an assemblage, by increasing the degree of internal homogeneity or the degree of sharpness of its boundaries, or destabilize it. The former are referred to as processes of *territorialization* and the latter as processes of *deterritorialization*.

Processes of territorialisation and deterritorialisation involve both spatial relations that define (or blur) specific territorial designations as well as non-spatial processes that increase or decrease the homogeneity within an assemblage – another way to understand this is to think of processes that stabilise or destabilise the assemblage. Along this dimension we can see stabilising structures that enable targeted killing like the laws of war as well as potential destabilising elements such as understandings of violence that denote subject positions that are highly gendered. As will be shown in subsequent chapters, acts of targeted killing, those who commit them, their component sensory regimes, and equipment used are embedded in prevailing notions of masculinity and femininity.

DeLanda (2006: 19) proposes that there is a third dimension to assemblages in which:

> specialized expressive media intervene, processes which consolidate and rigidify the identity of the assemblage or, on the contrary, allow the assemblage a certain latitude for more flexible operation while benefitting from...resources (processes of coding and decoding).

Within the targeted killing assemblage, this is the dimension occupied by culture and problematisations it generates. These provide the necessary mass to keep the assemblage in place, despite the occurrence of events that have the potential to destabilise the assemblage and reconfigure how political violence is deployed by liberal regimes.

For this project, the concept of the 'assemblage offers an emphasis on...the relations between history and potential, that is, of the different processes that historically produce...[targeted killing] and the possibilities for those conditions... to be contested, imagined differently and altered' (McFarlane 2011: 376). Where I differ from more rigid analyses of assemblages is in the following ways. First, I have not provided a total mapping of the micro-networks involved in the targeted killing assemblage. While my ontology and scope for analysis may be broader than many analysts of political violence would like, I have focused on elements of the targeted killing assemblage that I believe to be most important based on the role they play in shaping key problematisations that catalyse its production. The choice has been made because I wish to go beyond listing components as though their roles and relations are self-evident (Tonkiss 2011: 585). Second, I have not necessarily emphasised the agential qualities of objects to the extent common in analyses of assemblages. In part, this is because I see these agential qualities and their significance as always being mediated by culture and discourse; they are always also representations whatever else they may be (Lundborg and Vaughan-Williams 2015). Third, my primary interest is not just in the functional relations amongst component parts that comprise the targeted killing assemblage per se, but also in exploring the power dynamics underpinning them. Thus, my conceptualisation and framing of assemblage is one that emphasises the relationality of component parts, the incorporation of disparate elements (including the non-human), their power dynamics, and the plasticity of assemblages in so far as change is possible but not inevitable (i.e., an acknowledgement that process repetitions remain stable until minute changes – that may only be perceivable in hindsight – open the possibility for transformation). In the ontological balance between appearance and being, my account is more representational in emphasis. Thus, my view is one that reverses DeLanda's (2006: 2) dictum regarding the problem for a realist social ontology. My issue is not that material autonomy overwhelmingly determines referents, placing them outside of discourse, but that this only happens in rare cases and always in the context of institutions and practices that cannot be reduced simply to their material capacities or capabilities.[5]

As such, this analysis finds itself at the intersection of three of the most important movements in social theory of the past two decades: Foucauldian explorations of pastoralism, biopolitics, and liberalism; explorations of the law, the sovereign exception, and the limits of liberalism; and the new materialisms that have emerged from actor-network theory and object-oriented ontologies. While drawing upon relevant aspects from these bodies of research to advance the specific arguments that follow, it must be noted that at the heart of the analysis is also a friendly critique of them. Primarily, the critique centres on the way in which these forms of analysis have too often been deployed within the study of security such that they end up embracing the forms of technical rationalism that they seek to undermine. As will be shown, an awareness of the role of culture in the production of rationalisations, calculations of the capabilities of objects, and the conduits through which affects materialise is central to understanding the targeted killing assemblage.

My account is one that focuses more on what could be described as conditions of possibility shaping *ad bellum* considerations pertinent to targeted killing rather than seeking to specifically unpack the intricacies of specific *in bello* acts. This choice reflects two strategic decisions. The first is that while the political is never absent from the battle-space, the battle-space itself is constructed, produced, and maintained through broader political discourses. It is thus important to map how seemingly diverse sets of problematisations work together to create the conditions of possibility for particular forms of violence. The second is that while *in bello* analyses can directly speak to the mechanisms and processes involved in specific acts of killing in order to expose the gap between claims (of precision, proportionality, and effectiveness) and outcomes, this is achieved at the expense of examining where purported claims come from and how they have arisen. That the results of targeted killing operations rarely reflect stated claims about what targeted killing outcomes should be can provide important come-back but it risks diverting discussion towards technical issues (i.e., how to improve precision, proportionality, and effectiveness). Thus my preference is to focus on how particular desires and justifications are produced within the targeted killing assemblage. As a result, I have devoted less analysis to the technical specifics of the kill chain, detailed schematics of targeted killing infrastructures in battle-spaces, or the necropolitics of specific targeted killing events (e.g., Allinson 2015), but rather have attempted to capture the political *mis en scène* in which these become possible.

The data for capturing the political *mis en scène* has been compiled both through the graft, intuition, and blind luck of the lone scholar model of research as well as through crowd sourcing enabled by new forms of social media that arose in prominence during the process of researching and writing this book. Twitter feeds and blogs, in particular, made me aware of materials that might have otherwise escaped my attention and/or brought them to my attention more quickly than would have been possible by working in isolation. Sources of information for discerning key problematisations, and for the purposes of inferring the cultural dimensions of these problematisations, have encompassed primary governmental materials (including both approved/official reports and speeches as well as leaked classified documents), non-governmental research, media reports, published interviews, military memoires, and publicly available datasets, as well as academic literature across several fields. Given the interplay of public prominence and secrecy within the assemblage itself where individual acts of killing are publicly acknowledged but the processes enabling specific acts of killing may be left more opaque, I have erred on the side of caution when presenting claims, seeking either confirmation through multiple sources, or preferably triangulation across different types of sources. This has at times led to a mixed method approach whereby different forms of data capture and analysis have been mobilised (Fielding 2012). The overall ethos of inquiry has been driven by two insights into critical method identified by Claudia Aradau and Jef Huysmans (2014). The first is their contention that methods are performative practices that assemble epistemologies, ontologies, theories, concepts, and data rather than representing a direct means of aligning knowledge production with reality (Aradau and Huysmans 2014: 598). The second is that any method has the potential to be disruptive to relations of power and it is the purposes to which methods are used that are important in this regard (Aradau and Huysmans 2014: 609). Further discussion appears in subsequent chapters when necessary.

My research approach has raised two ethical issues. The first was the ethics of using grey materials (i.e., leaked classified documents). My position was pragmatic rather than absolutist (Jenkins 2013). The materials I consulted (and in some cases cited) contained information that I believe is in the global public interest. I am unconvinced that making public the rules of engagement for counter-insurgency in specific war theatres, how persons of interest are being tracked within the 'kill chain', or the ways in which targeted killing is being undertaken causes direct harms full stop, let alone in light of the benefits accrued by broader publics when there is a better collective understanding of how global counter-insurgency works in practice. The second is the ethics of crowd sourcing research materials via social media. Most literature on the ethics of crowd sourcing in research focuses on systematic recruitment of members of the public to conduct analyses of large datasets that require time-consuming forms of examination (e.g., mapping strands of DNA), often facilitated by turning the process into a 'game' (Graber and Graber 2013). With regards to data collection and crowd sourcing via social media, key issues raised within this literature, such as covertly transforming participants into experimental subjects, are not relevant concerns.[6] Primarily, practices of sharing were being initiated by those who were already on social media and these were being undertaken independently of my own research; in other words, I was not inviting others to undertake a task that they were not already doing of their own volition for the purposes of activism, scholarly communication, and dissemination. Moreover, I sought to contribute to this ecosystem myself by sharing links, commentaries, observations, and more formal papers, ensuring that the contributions of others were being reciprocated to the best of my own ability.

In sum, I have deployed a methodological position, conceptual apparatus, theoretical outlook, and methods that help to get at the question of how targeted killing by liberal regimes has become possible. The analysis that follows will demonstrate that a greater sensitivity to cultural dynamics is required to reach a richer understanding of how targeted killing has been positioned as legitimate security practice within regimes that are ostensibly liberal in governing outlook. By being sensitive to culture, one can incorporate an important element in the catalytic dynamics of targeted killing that contributes to the contingently obligatory relations that mobilise capabilities towards this violent practice. In the next section I outline how problematisations and styles of thought can shape these contingently obligatory relations.

Problematisations and styles of thought

In the mobilisation by liberal regimes of the exercise of capabilities for targeted killing, how it is being thought about (i.e., problematised) within the political, legal, and security realms is shaped by a constellation of elements. Some of these reflect longer-standing ambiguities over the meaning of key concepts germane to questions of global order including sovereignty, territoriality, and terrorism. Others such as risk, technology, and the influence of gender and racial representations of the act itself are historically embedded, as well as being greatly conditioned by the current discursive context of contemporary liberalism and its constitutive relations of power. These problematisations are not merely academic. Nor do they operate in isolation. Rather, problematisations shape the ways in which actors – either state or non-state – approach questions of assassination and targeted killing. Problematisations are a vital area for political study. They reveal the ways in which people, places, things, and issues are constructed and framed as a particular type of problem as opposed to some other. A problematisation relies on forms of knowledge, common sense thinking, the limits of intelligible discourse, and cultural predispositions to shape how the 'problem' is understood. Often in contemporary politics, the way in which problems are defined by elites, the media, and other centres of authority are taken at face value by myriad actors that make up civil society.

The importance of analysing problematisations lies in how they may be accepted by disparate actors who may agree on very little else. How a problem is defined will greatly condition the way in which the problem is addressed, managed, or even ignored. Moreover, problematisations are never a product of neutral arbitration or disinterested analysis; at their core, they are constitutive of more generalised relations of political, social, economic, and cultural power. Conversely, power-relations themselves play a distinctive role in determining particular problematisations. Their channels, circuits, and flows shape how issues get defined. And in these processes of politicisation and contestation, sometimes the problem is the problem itself – that is, the problem is identified in a way that limits the scope for rethinking, revision, and reconfiguration.

In thinking of problematisations and power-relations, Nikolas Rose (2007) argues that Ludwik Fleck's concept of a 'style of thought' is useful. Rose (2007: 12) outlines that:

> a style of thought is a particular way of thinking, seeing, and practicing. It involves formulating statements that are only possible and intelligible within that way of thinking. Elements – terms, concepts, assertions, references, relations – are organized into configurations of a certain form that count as arguments and explanations. Phenomena are classified and sorted according to criteria of significance. Certain things are designated as evidence and used in certain ways...And of course, a style of thought in an area...also embodies a way of identifying difficulties, questioning arguments, identifying explanatory failures – a mode of criticism, of error seeking and error correction.

Thus, the components that contribute to a specific style of thought are an essential element in shaping how issues become problematised including how they are imagined, how their various dimensions are assembled into a whole, and the types of knowledge production that are used to find 'solutions' to their perceived effects or to 'manage' them. And the terrains of imagination, assemblage, knowledge production, and policy are all sites in which power operates.[7]

Central to these practices of terraforming are processes of objectification and subjectification that 'organize the diagnostic, forensic, and interpretive gaze[s]' of authorities (Rose 2007: 140). These classifications both divide and unify by delineating who is to be managed in particular ways – in the practices of security, counter-insurgency, and law – as these delineations themselves gloss over marked differences that may be contained within specific classifications. Thus, categories based on contingent and ultimately often unsustainable distinctions such as civilian, terrorist, enemy combatant, irregular forces, security threat, criminal, and person of interest actually begin to shape the production of truth, not only in terms of the analytic techniques used, but also in terms of the kind of data that is collected and analysed and the way in which differences are found. Thus, Rose (2007: 174) reminds us that 'classifications that may be arbitrary and contingent – "political constructs" – are made real in the very process of using them within a technology of investigation and analysis'.

In focusing on the ways in which targeted killing is being problematised within its assemblage and how specific types of problematisation become possible, the role of political context – as defined by relations of power – is paramount. While it is often assumed that the realms of law, science, and military strategy are technical enterprises that are separate from politics – apart from 'abnormal' circumstances in which they become intertwined – the ways in which these realms frame problems and then act upon them is highly political.[8] Law, science, and military strategy construct the (structural) conditions under which targeted killing is contested and practiced. These conditions can be constituted both through discursive and material forces that frame how issues are problematised as well as the limits of what is considered to be practical and/or politically possible. The contemporary war on terror is a prime example of a problematisation that has infiltrated thinking about a wide range of issues from the architectural design of public spaces to practices of societal integration.

Examining problematisations is also useful in that the concurrent concern with practice and power necessarily places a demand on the analyst to focus on the role of contingency in their construction. Problematisations are neither natural outcomes based on a teleological progression in modes of thinking nor are they solidly fixed in historical time and place. They are fluid, always in a state of flux, and a reflection of broader contingencies. Therefore, the role of the analyst in an examination of problematisations is to bring a focus to the contingencies at play. In tracing the contingency of relations of power, problematisations, and practices, Rose (2007: 5) – via Michel Foucault – argues that we construct genealogies of the present, a mode of analysis that 'in making…contingencies thinkable, in tracing the heterogeneous pathways that led to the apparent solidity of the present, in historicizing those aspects of our lives that appear to be outside of history, in showing the role of thought in making up our present…[seeks] to make that present open to reshaping.'

At the same time, although being able to demonstrate that the present is a contingent condition is important, it is also important not to lose sight that the future – despite our current perceptions of a rigid spectrum of possibility predominantly defined by the pessimisms and worst case scenarios engendered by what Derek Gregory (2011) has called the 'everywhere war' – is also open to radical rethinking. By 'demonstrating that no single future is written in our present, it might fortify our abilities, in part through thought itself, to intervene in that present, and so to shape something of the future that we might inhabit' (Rose 2007: 5). Thus, in exposing the targeted killing assemblage, deterritorialising lines of flight can be revealed.

The cultural politics of targeted killing: lines of flight

The argument that will be presented in the chapters that follow will examine how targeted killing (and assassination as its other) are currently conceptualised along five lines of flight that contribute to the cultural politics of targeted killing. The first line of flight is the law. Within liberal regimes, assassination and targeted killing are positioned in relation to the laws of war and human rights. It is also within the *mise en scéne* of the law that debates regarding who can be legitimately targeted and under what conditions this is permissible take place. While the central questions framing the legal problematic is whether targeted killing is legal, it raises broader questions regarding liberalism, the exception as normative suspension, and the role of cultural politics in shaping what is permissible.

The second line is the political dimension in which assassinations/targeted killings are identified as events, contestations occur over meaning, and efforts are undertaken to convince others of the (il)legitimacy of a course of action in particular circumstances. The question that best defines the political pro-blematisation is a relatively simple one: 'who may be assassinated or targeted'? Yet the simplicity is deceptive. The ways in which this question has been answered have been formed in interaction with complex representations and culturally contingent understandings of the rituals of power and resistance.

The third line is the nexus formed by strategic understandings of the con-temporary battle-space as well as predominant views of the roles of networks and adaptive change within contemporary warfare. It will be demonstrated how these are productive of a political economy that shares the same value system, enabling military transformation, the commodification of the drone, and the capacity for particular forms of targeted killing.

The fourth line is made up of the aesthetic subjects produced through targeted killing in drone warfare. With drone strikes premised on a belief in the calculability of (re)actions of human and machine, as well as the possibility of total control over these elements, it will be shown how drones, operators, and targeted populations exceed these parameters. Thus, even when culturally derived, the experiences and forms of agency enabled by targeted killing extend beyond what can currently be acknowledged by liberal authorities.

The fifth line is the quotidian geopolitics underpinning drone strikes. Where existing research has predominantly centred on the extraordinary or exceptional spaces engendered by drone warfare, the focus here is on how this form of political violence affects everyday places. The primary unit to be examined is the home and how the sacredness of the home does not register in the geographical imagination informing counter-insurgency doctrine.

The final line of flight brings together relationships and themes that run across the targeted killing assemblage including gender, ways of seeing, power dynamics, and cultural predispositions. The argument presented is that these do not just reveal how targeted killing has become possible in contemporary counter-insurgency. More importantly, they demonstrate the centrality of violence to liberal forms of rule.

Notes

1 This volume of civilian deaths resulting from attempts to eliminate a 'high value' target has proven not to be unusual. *The Guardian* would later report in 2014 that

data analysis undertaken by Reprieve, a human rights organisation, showed that targeted killing operations which sought to eliminate 41 named individuals had directly contributed to the deaths of 1,147 people (Ackerman 2014).

2 While drone is the word used in common parlance, RPA is the preferred term in military circles. This preference is based on the fact that these platforms *are* piloted and not autonomous systems (as would be implied by the term 'drone'). However, both will be used interchangeably throughout the monograph. This is primarily a stylistic choice to avoid repetition.

3 The nomenclature of assassination though has transformed since the times in which its significance was related to regicide or tyrannicide. The contemporary cultural politics of targeted killing is increasingly imbricated within the cultural geopolitics of the drone. In being what Levi Bryant (2014: 202–205) would refer to as a 'bright object', the drone has significantly shaped the cultural politics of targeted killing since the turn of the past century. At the same time, as will be demonstrated in chapter 3, the legacies of the pre-drone era still emit a powerful gravitational pull.

4 For more on the general politics of 'precision' weapons systems and legitimacy, see Beier (2003).

5 This is not to claim that social realists ignore language. For example, DeLanda (2006: 3) notes that 'language plays an important but not a constitutive role' in the construction of assemblages. My position differs in that I am unwilling to cede that an assemblage, its components, dynamics, and effects have any political importance outside of their discursive conditions of emergence while conceding that the properties and capabilities do not necessarily require forms of representation to 'exist'.

6 Beyond formal publishing distribution networks, academics have shared and dis- seminated research materials through various mechanisms for decades. The shift over the past 10 years has been how social media platforms such as Twitter, Academia.edu, and Rich Site Summary (RSS) feed applications (e.g., Feedly) have helped to systematise the ways in which these materials can be distributed and collected.

7 As Rose (2007: 12) eloquently notes, a style of thought also 'shapes and establishes the very object of explanation, the set of problems, issues, phenomena that any problematisation and its resulting explanations attempt to answer'.

8 In fact, these disciplines have been greatly shaped by the problematisations they have been set forth to solve.

Bibliography

Ackerman, S. (2014) '41 Men Targeted but 1,147 People Killed: US Drone Strikes – The Facts on the Ground'. *The Guardian*. Available from: http://www.theguardian.com/us-news/2014/nov/24/-sp-us-drone-strikes-kill-1147. Accessed 15 October 2015.

Adams, J. (2010) 'US Defends Unmanned Drone Attacks after Harsh UN Report'. *Christian Science Monitor*. Available from: http://www.csmonitor.com/World/terror ism-security/2010/0603/US-defends-unmanned-drone-attacks-after-harsh-UN-report. Accessed 15 October 2015.

Agamben, G. (1998) *Homo Sacer: Sovereign Power and Bare Life*, Stanford, Stanford University Press.

Allinson, J. (2015) 'The Necropolitics of Drones', *International Political Sociology*, 9(2), 113–127.

Alston, P. (2010) *Addendum to the Report of the Special Rapporteur on extrajudicial, summary, or arbitrary executions: Study on targeted killings*, New York, Human Rights Council, United Nations.

Aradau, C. (2010) 'Security That Matters: Critical Infrastructure and Objects of Protection', *Security Dialogue*, 41(5), 491–514.

Aradau, C. and Huysmans, J. (2014) 'Critical Methods in International Relations: The Politics of Techniques, Devices and Acts', *European Journal of International Relations*, 20(3), 596–619.

Aradau, C. and van Munster, Rens (2009) 'Exceptionalism and "The War on Terror"': Criminology Meets International Relations', *British Journal of Criminology*, 49(5), 686–701.

Bazan, E. B. (2002) *Assassination Ban and E.O. 12333: A Brief Summary*, Washington, Congressional Research Service, The Library of Congress.

BeierJ. M. (2003) 'Discriminating Tastes: "Smart" Bombs, Non-combatants, and Notions of Legitimacy in Warfare, *Security Dialogue*, 34(4), 411–425.

Bell, J. B. (2005) *Assassin: Theory and Practice of Political Violence*, New Brunswick, Tranaction Publishers.

Ben-Yehuda, N. (1990) 'Gathering Dark Secrets, Hidden and Dirty Information: Some Methodological Notes on Studying Political Assassinations', *Qualitative Sociology*, 13(4), 345–371.

Ben-Yehuda, N. (1997) 'Political Assassination Events as a Cross-cultural Form of Alternative Justice', *International Journal of Comparative Sociology*, 38(1–2), 25–47.

Bryant, L. R. (2014) *Onto-cartography: An Ontology of Machines and Media*, Edinburgh, Edinburgh University Press.

Bureau of Investigative Journalism (2011) 'Obama 2009 Pakistan Strikes', *Bureau of Investigative Journalism*. Available from: https://www.thebureauinvestigates.com/ 2011/08/10/obama-2009-strikes. Accessed 15 October 2015.

Byman, D. (2006) 'Do Targeted Killings Work?', *Foreign Affairs*, 85(2), 95–111.

Campbell, D. (1998) *Writing Security: United States Foreign Policy and the Politics of Identity*, Minneapolis, University of Minnesota Press.

Carvin, S. (2012) 'The Trouble with Targeted Killing', *Security Studies*, 21(3), 529–555.

Carvin, S. and Williams, M. J. (2015) *Law, Science, Liberalism, and the American Way of Warfare*, Cambridge, Cambridge University Press.

Chamayou, G. (2015) *Drone Theory*, New York, Penguin.

Closs Stephens, A. and Vaughan-Williams, N. (eds) (2008) *Terrorism and the Politics of Response: London in a Time of Terror*, London, Routledge.

Cloud, D. S. (2010) 'U.N. Report Faults Prolific Use of Drone Strikes by U.S. Los Angeles Times '. Available at: http://articles.latimes.com/2010/jun/03/world/la-fg-cia -drones-20100603. Accessed 15 October 2015.

Dauphinee, E. and Masters, C. (ed) (2007) *The Logics of Biopower and the War on Terror: Living, Dying, Surviving*, New York, Palgrave-Macmillan.

Dean, M. (1999) *Governmentality: Power and Rule in Modern Society*, London, Sage.

DeLanda, M. (2006) *A New Philosophy of Society: Assemblage Theory and Social Complexity*, London, Bloomsbury Publishing.

Deleuze, G. and Guattari, F. (1987) *A Thousand Plateaus: Capitalism and Schizophrenia* (B. Massumi, Trans.), Minneapolis, University of Minnesota Press.

Der Derian, J. (1995) 'The Value of Security: Hobbes, Marx, Neitzsche, and Baudrillard' in R. Lipschutz (ed.), *On Security*, New York, Columbia, 24–45.

Dillon, M. and Reid, J. (2009) *The Liberal Way of Warfare: Killing to Make Life Live*, London, Routledge.

de Larrinaga, M. and Doucet, M. (2008) 'Sovereign Power and the Biopolitics of Human Security', *Security Dialogue*, 39(5), 517–537.

Downes, C. (2004) '"Targeted Killings" in an Age of Terror: The Legality of the Yemen Strike', *Journal of Conflict and Security Law*, 9(2), 277–294.

Duffield, M. (2007) *Development, Security, and Unending War: Governing the World of Peoples*, Cambridge, Polity Press.

Edkins, Veronique J.Pin-Fat and Michael J. Shapiro (ed.) (2004) *Sovereign Lives: Power in Global Politics*, New York, Routledge.

Enemark, C. (2013) *Armed Drones and the Ethics of War: Military Virtue in a Post-heroic Age*, Abingdon, Routledge.

Falk, O. (2014) 'Permissibility of Targeted Killing', *Studies in Conflict & Terrorism*, 37(4), 295–321.

Fielding, N. G. (2012) 'Triangulation and Mixed Methods Designs Data Integration With New Research Technologies', *Journal of Mixed Methods Research*, 6(2), 124–136.

Finkelstein, C., Ohlin, J. D. and Altman, A. (2012) *Targeted Killings: Law and Morality in an Asymmetrical World*, Oxford: Oxford University Press.

Fletcher, G. P. (2006) 'The Indefinable Concept of Terrorism', *Journal of International Criminal Justice*, 4(5), 894–911.

Ford, F. L. (1987). *Political Murder: From Tyrannicide to Terrorism.* Cambridge, Harvard University Press.

Foucault, M. (2003) *Society Must Be Defended: Lectures at the College de France 1975–1976*, New York: Picador.

Foucault, M. (2007) *Security, Territory, Population: Lectures at the College de France 1977–1978*, Basingstoke, Palgrave.

Freeman, M. (2014) 'A Theory of Terrorist Leadership (and its Consequences for Leadership Targeting)', *Terrorism and Political Violence*, 26(4), 666–2212;687.

Gordon, A. (2006) 'Purity of Arms, Preemptive War, and Selective Targeting in the Context of Terrorism: General, Conceptual, and Legal Analyses', *Studies in Conflict and Terrorism*, 29(5), 493–508.

Graber, M. A. and Graber, A. (2013) 'Internet-based Crowdsourcing and Research Ethics: The Case for IRB Review', *Journal of Medical Ethics*, 39(2), 115–118.

Grayson, K. (2008) 'Human Security as Power/Knowledge: The Biopolitics of a Definitional Debate', *Cambridge Review of International Affairs*, 21(3), 383–401.

Gregory, D. (2011) 'The Everywhere War', *The Geographical Journal*, 177(3), 238–250.

Gregory, T. (2015) 'Drones, Targeted Killings, and the Limitations of International Law', *International Political Sociology*, 9(3), 197–212.

Grusin, R. (2010) *Premediation: Affect and Mediality after 9/11*, Basingstoke, Palgrave Macmillan.

Guiora, A. (2004) 'Targeted Killing as Active Self-Defense', *Case Western Reserve Journal of International Law*, 36(1), 319–334.

Hafez, M. M. and Hatfield, J. M. (2006) 'Do Targeted Assassinations Work? A Multivariate Analysis of Israel's Controversial Tactic during Al-Aqsa Uprising', *Studies in Conflict and Terrorism*, 29(4), 359–382.

Hall, S. (1997) *Representation: Cultural Representations and Signifying Practices*, London, Sage.

Hallissy, M. (1987) *Venomous Women*, New York, Greenwood Press.

Hepworth, D. P. (2014) 'Terrorist Retaliation? An Analysis of Terrorist Attacks Following the Targeted Killing of Top-tier al Qaeda Leadership', *Journal of Policing, Intelligence and Counter Terrorism*, 9(1), 1–18.

Honig, O. (2007) 'Explaining Israel's Misuse of Strategic Assassinations', *Studies in Conflict and Terrorism*, 30(6), 563–577.

Iqbal, Z. and Zorn, C. (2006) 'Sic Semper Tyrannis? Power, Repression, and Assassination Since the Second World War', *The Journal of Politics*, 68(3), 489–501.

Jenkins, P. (2013) 'Absolutist, Pragmatist and Realist Approaches to Research Ethics in the Digital Humanities: The Case of the Schneerson Collection'. Available from: http://papers.ssrn.com/sol3/papers.cfm?abstract_id=2474183. Accessed 15 October 2015.

Johnston, P. B. (2012) 'Does Decapitation Work? Assessing the Effectiveness of Leadership Targeting in Counterinsurgency Campaigns', *International Security*, 36(4), 47–79.

Jordan, J. (2009) 'When Heads Roll: Assessing the Effectiveness of Leadership Decapitation', *Security Studies*, 18(4), 719–755.

Jordan, J. (2014) 'Attacking the Leader, Missing the Mark: Why Terrorist Groups Survive Decapitation Strikes', *International Security*, 38(4), 7–38.

Kaag, J. and Kreps, S. (2014) *Drone Warfare*, Cambridge, Polity Press.

Kasher, A. and Yadlin, A. (2005) 'Assassination and Preventive Killing', *SAIS Review*, 25(1), 41–57.

Kessler, O. and Werner, W. (2008) 'Extrajudicial Killing as Risk Management', *Security Dialogue*, 39(2–3), 289–308.

Kirkham, J. F., Levy, S. G., and Crotty, W. J. (1970). *Assassination and Political Violence: A Report*, New York, Praeger.

Kober, A. (2007) 'Targeted Killing during the Second Intifada: The Quest for Effectiveness', *Journal of Conflict Studies*, 27(1), 76–93.

Kretzmer, D. (2005) 'Targeted Killing of Suspected Terrorists: Extra-Judicial Executions or Legitimate Means of Defence?', *European Journal of International Law*, 16(2), 171–212.

Lobo-Guerrero, L. (2007) 'Biopolitics of Specialized Risk: An Analysis of Kidnap and Ransom Insurance', *Security Dialogue*, 38(3), 315–334.

Lobo-Guerrero, L. and Dillon, M. (2008) 'Biopolitics of Security in the 21st Century', *Review of International Studies*, 34(2), 265–292.

Lundborg, T. and Vaughan-Williams, N. (2015) 'New Materialisms, Discourse Analysis, and International Relations: A Radical Intertextual Approach', *Review of International Studies*, 41(1), 3–25.

Massumi, B. (2007) 'Potential Politics and the Primacy of Preemption', *Theory and Event*, 10(2).

Masters, J. (2013) 'Targeted Killings', *Council on Foreign Relations*. Available from: http://www.cfr.org/counterterrorism/targeted-killings/p9627. Accessed 15 October 2015.

McCrisken, T. (2013) 'Obama's Drone War', *Survival: Global Politics and Strategy*, 55(2), 97–122.

McFarlane, C. (2011) 'On Context: Assemblage, Political Economy, and Structure', *City*, 15(3–4), 375–388.

Miller, P. and Rose, N. (2008) *Governing the Present*, Cambridge, Polity Press.

Neal, A. W. (2006) 'Foucault in Guantánamo: Towards an Archaeology of the Exception', *Security Dialogue*, 37(1), 31–46.

Neocleous, M. (2008) *Critique of Security*, Edinburgh, University of Edinburgh Press.

Neocleous, M. (2011) '"A Brighter and Nicer New Life": Security as Pacification', *Social & Legal Studies* 20(2): 191–208.

Neocleous, M. (2013) 'The Dream of Pacification: Accumulation, Class War, and the Hunt', *Socialist Studies/Études socialistes*, 9(2), 7–31.

O'Brien, K. (2001) 'The Use of Assassination as a Tool of State Policy: South Africa's Counter-Revolutionary Strategy 1979–1992(Part II)', *Terrorism and Political Violence*, 13(2), 107–142.

Protevi, J. (2009) *Political Affect: Connecting the Social and the Somatic*, Minneapolis, University of Minnesota Press. Kindle Edition.

Protevi, J. (2010) 'Rhythm and Cadence, Frenzy and March', *Theory and Event*, 13(3).

Price, R. (1997) *The Chemical Weapons Taboo*, Ithaca: Cornell University Press.

Puar, J. K. (2007) *Terrorist assemblages: Homonationalism in Queer Times*, Durham, Duke University Press.

Rabinow, P. and Rose, N. (2003) 'Introduction: Foucault Today' in P. Rabinow and N. Rose (eds), *The Essential Foucault: Selections from Essential Works of Foucault 1954–1984*. New York, The New Press, vii–xxxv.

Rae, J. D. (2014) *Analyzing the Drone Debates: Targeted Killings, Remote Warfare, and Military Technology*, New York, Palgrave Macmillan.

Rogers, A. and Hill, J. (2014) *Unmanned: Drone Warfare and Global Security*, London, Pluto Press.

Rose, N. (2007) *The Politics of Life Itself: Biomedicine, Power, and Subjectivity in the Twenty-first Century*, Princeton, Princeton University Press.

Salter, M. (2006) 'The Global Visa Regime and the Political Technologies of the International Self: Borders, Bodies, Biopolitics', *Alternatives*, 31(2), 167–189.

Schwarz, E. (2015) 'Prescription Drones: On the Techno-biopolitical Regimes of Contemporary "Ethical Killing"', *Security Dialogue*, 47(1), 59–75.

Sheppard, E. (2008) 'Geographic Dialectics?', *Environment and Planning A*, 40(11), 2603–2612.

Sloggett, D. (2014) *Drone Warfare: The Development of Unmanned Aerial Conflict*, Barnsley, Pen and Sword.

Statman, D. (2004) 'Targeted Killing', *Theoretical Inquiries in Law*, 5(1), 179–198.

Stein, Y. (2003) 'Response to Israel's Policy of Targeted Killing: By Any Name Illegal and Immoral', *Ethics & International Affairs*, 17(1), 127–137.

Stocker, M. (1998) *Judith Sexual Warrior: Women and Power in Western Culture*, New Haven and London, Yale University Press.

Strawser, B. J. (ed.) (2013) *Killing by Remote Control: The Ethics of an Unmanned Military*, Oxford, Oxford University Press.

Thomas, W. (2000) 'Norms and Security: The Case of Assassination', *International Security*, 25(1), 105–133.

Tonkiss, F. (2011) 'Template Urbanism', *City*, 15(5), 584–588.

Tuastad, D. (2003) 'Neo-Orientalism and the New Barbarism Thesis: Aspects of Symbolic Violence in the Middle East Conflict(s)', *Third World Quarterly*, 24(4), 591–599.

Weldes, J. MarkLaffey, HughGusterson, and Raymond Duvall (1999) 'Introduction: Constructing Insecurity', in *Cultures of Insecurity: States, Communities, and the Production of Danger*, Minneapolis, University of Minnesota Press, 1–33.

Williams, B. G. (2013) Predators:*The CIA's Drone War on Al Qaeda*, Washington DC, Potomac Books, Inc.

Williams, R. (1985) *Keywords: A Vocabulary of Culture and Society*, Oxford, Oxford University Press.

Wilner, A. S. (2010) 'Targeted Killings in Afghanistan: Measuring Coercion and Deterrence in Counterterrorism and Counterinsurgency', *Studies in Conflict & Terrorism*, 33(4), 307–329.

Zenko, M. (2014) 'America's 500th Drone Strike. Politics, Power, and Preventative Action', *Council on Foreign Relations*. Available from http://blogs.cfr.org/zenko/2014/11/21/americas-500th-drone-strike/#. Accessed 15 October 2015.

Zussman, A. and Zussman, N. (2006) 'Assassinations: Evaluating the Effectiveness of an Israeli Counterterrorism Policy Using Stock Market Data', *Journal of Economic Perspectives*, 20(2), 1–15.

2 Beyond the exception
The legal problematisation of targeted killing

Introduction

Within the targeting killing assemblage, the legal sphere has become a primary location for both the problematisation of targeted killing and for its contestation. In essence, it has become a site of lawfare, that is, for the use of legal means to secure strategic geopolitical objectives through the production of specific forms of governance (e.g., Morrissey 2011; Jones 2015b). Within the assemblage, it is bodies of law and legal processes that have helped to stabilise the identity of targeted killing, not just by casting it in opposition to the lawlessness of assassination, but also by catalysing processes that produce spaces for its legal commissioning. But what must be emphasised is that in the case of targeted killing – and drone strikes – permissibility is not made possible by the sovereign exception in which the symbolic normativity of the law is suspended until such a time that the existing legal order – or a new legal order – can (re)form (Huysmans 2006b). Rather, these spaces come to fruition through the normal workings of the law and law-making procedures (Neal 2012). Identifying the exception – as a normative suspension – has been used by critics (e.g., Franck 2004; Kramer and Michalowski 2005; Welch 2007) to explain how it becomes possible to monitor, detain, torture, pre-emptively attack and kill under the auspices of global counter-insurgency. Unfortunately, while rhetorically powerful, the normative suspension argument can be both politically naïve and unhelpful. It can be naïve insofar as it accepts liberalism's own self-legitimating discourse regarding the limited coercive power of the liberal state and adherence to the principles of humanitarian warfare as the baseline for determining the extraordinariness of acts of violence; it is an acceptance that the law is separate from violence and that the law instantiates justice, in procedural form, if not in substance. It can also be unhelpful insofar as this naivety then misdirects criticism away from a considerable source of power for the commissioning of violence by liberal states: the law.

As such, a critique that does not take the exception as given requires a careful rethinking of how targeted killing is positioned under the law. The question then becomes how targeted killing works within the law rather than violating it from first principles. Such a critique provokes a more disturbing reassessment of the relationship constituted by violence and the liberal state as well as the law and

justice under liberalism. It foregrounds the attempt to impose 'a set of substantive values' in global counter-insurgency by claiming the 'civilizational
higher ground' (Huysmans 2006a: 26) These conclusions are in part guided by
Jacques Derrida's (1989/1990) reading of Walter Benjamin in 'The Force of Law'
as well as an engagement with the political theology of Carl Schmitt. At the
same time, it suggests that critiquing targeted killing solely in regards to formal
relationships constituted by politics and the law does not do justice to the
political, cultural, economic, and spatial enablers of political violence. The conceptual elements of the argument for this chapter proceed in three parts. First, I
will outline the relationship between the law and violence in general terms. I will
then outline the case made for the exception as normative suspension in criticisms of contemporary violence. I then turn to Schmitt's own criteria for the
exception as well as Giorgio Agamben's (1998) reformulation to provide an
evaluative framework for the empirical work on the law to follow.

Empirically, how targeted killing is positioned within international humanitarian law and international human rights law will then examined. Three
findings demonstrate how the law and legal discourses stabilise targeted killing.
The first is a legal distinction made between assassination – which is illegal by
first principles under international humanitarian law – and targeted killing. The
second is the identification of four concepts that are shaping problematisations of
targeted killing under *jus ad bellum* and *jus in bello* considerations: pre-emption
(or what is sometimes called anticipatory self-defence), proportionality, sovereignty, and territoriality.[1] These concepts create space for targeted killing while
enabling its commission. Therefore, the discursive shift from assassination to
targeted killing – and the concurrent move to targeted killing via RPA strikes – is
important.[2] Rather than representing mere changes in terminology or tactics,
they are processes that contribute to the territorialisation of the targeted killing
assemblage. The third is to show how international human rights law contains
space for the commissioning of targeted killing.

The analysis will then move into a consideration of the legal rationales
currently used by the United States and Israel to justify their use of targeted
killing. Their selection rests on two primary criteria. First, both have utilised
targeted killing as a part of their counter-insurgency strategies. Second, both
have traditions of 'man-hunting' (Chamayou 2012) within their political
cultures. Third, both states understand themselves to be liberal democracies.
The point here is not to evaluate the legal merits of the justifications offered,
but rather to demonstrate that both the United States and Israel offer defences
for targeted killing that make explicit use of legal principles rather than
declaring their suspension. The implications of the relationship of the law and
targeted killing will then be discussed in the conclusion.[3]

Violence, the law, and the exception

In his discussion of law, Derrida (1989/1990: 927) – via Benjamin – argues
that its 'originary force' comes through violence. Whether this derives from

the metaphorical violence of defining and delimiting actions from the realm of un-decidability into a calculable matrix, or from the kinetic force that establishes a new political order that then reshapes the un-decidable into the calculable, is less important than the fact that there is a rupture, a coerced emergence that establishes new rules of conduct. This point of emergence and its authoritative legitimacy is a profoundly theological event. It is an event whose righteousness must be taken on as an article of faith: by definition, the origin of authority for law cannot rest on anything but the law itself. This means that the law is always premised on the potential deployment of violence without a ground – and its originary force can neither be legal or illegal: it just 'is' (Derrida 1989/1990: 943). The reading of the law offered by Derrida points to its unstable foundation, both in its genealogy and with regards to its legitimacy. The political implication is that the success of a legal system is not measured by its ability to provide justice but rather in its ability to produce 'interpretative models to read in return, to give sense, necessity and above all legitimacy, to the violence that has produced, among others, the interpretative model in question...' (Derrida 1989/1990: 993).

As a form of violence governed by legal codes and treatises, war is no different in this regard. Whereas there has been a tendency to view state violence of late as falling outside of the symbolic normative constraints of the law – perhaps in part, a side-effect of popular views of the second Gulf War – it bears reflecting, as will be shown below, that warlike violence is always deployed within the sphere of the law. As Derrida notes, the political implications of war's mutually constitutive relationship with law is not just that victory in war establishes new law – either through the suing of peace or violent practices that become established as convention. War is even more fundamental; it is not just originary 'violence in pursuit of natural ends... [but] is in fact a violence that serves to *found* law...' (Derrida 1989/1990: 999; italics added).

As will be shown below, the quasi-mystical origins of law and its inextricable links to violence do not produce a field free of contestation. Despite the protestations of legal positivists and those (reactionary) forces who wish to conceal the lack of an authoritative foundation, *the law is political*. It is political because it is both imbricated in the construction of the relations of power that define a particular order as well as being constituted by principles that are unstable in their interpretation and meaning. The contestation over principles, their applicability, their definition, and meaning are 'politicization[s that oblige]... one to reconsider and so to reinterpret the very foundations of law such as they had previously been calculated or delimited (Derrida 1989/1990: 971). This inherent instability and contingency, that is the congenital politicisation of the law, is both an opening for decisionism – as a reactionary remedy for the challenges of un-decidability – and the basis for lawfare, with each seeking to territorialise space for the legal commissioning of violence. While originally conceptualised as a form of asymmetrical combat undertaken by civil society actors to constrain the war-fighting capabilities of states, lawfare, its tactics, and *raison d'etre* have changed. Today, lawfare has been commandeered by states for the purposes of pacification. The law is thus used in tandem with other tactics to achieve core geostrategic objectives (Dunlap 2009: 35). The law is a supplement to war-fighting by acting as a 'force multiplier' (Morrissey 2011: 291). I have argued elsewhere that:

> Not only does the incorporation of legal frameworks attempt to provide an extra-strategic legitimating rationale for targeted killing, but the resort to the complexities of the law potentially de-politicises the practice by presenting its acceptability as a technical question for legal experts.
>
> (Grayson 2012: 122)

Thus, within these processes, the law has an interest in maintaining a monopoly on violence, both in its sanctioning and legal administration (Derrida 1989/1990: 985). But more importantly, what is often overlooked is the inverse relation: violence itself is interested in maintaining a monopoly on the law. The establishment of a framework in which violence becomes the regulatory norm, that is, where violence is conceived as a routine, uneventful, and rule governed act of legitimate authority, promotes both a mystification of the legal order as well as the transposition of violence into the realm of technical banality. It maintains a scope for the commissioning of violence by legal means through the invocation of a complex architecture of procedural decrees that then divests the commissioner of any liability for damages. Thus, Derrida (1989/1990: 925) argues that the law is a justifying force, even if it is judged from elsewhere to be unjustifiable.

The exception

The understanding of law above could be characterised as more cynical than most. It also does not reflect more commonly held views of the law as a constraint on state power within liberal systems of rule. Moreover, the close connection identified between the law and violence is unsettling. While the recognition of the split between the law and justice is at least as old as the death of Socrates, presenting violence undertaken by liberal states and the law as mutually constitutive is a radical shift from more orthodox liberal views of the law. Yet Derrida's reading of the law paradoxically parallels critiques of liberalism that have emerged through analyses of the politics of the exception. The views of Schmitt and Agamben are helpful here in so far as they establish thresholds through which we might discern the presence of exceptional politics as a means of understanding the relationship amongst targeted killing, the law, and sovereignty.

Schmitt (1985: 5) begins *Political Theology* with the statement that 'sovereign is he [sic] who decides on the exception'. But emergency measures in and of themselves do not necessarily meet Schmitt's threshold of the exception. He argues that:

> what characterises an exception is principally unlimited authority, which means the suspension of the entire existing [legal] order. In such a

situation, it is clear that the state remains whereas the law recedes. Because the exception is different from anarchy and chaos, order in the juristic sense still prevails even if it is not of the ordinary kind.

(Schmitt 1985: 12)

Sovereign power is central to Schmitt's (1985: 13) understanding of the exception because 'all law is "situational law"'. Any legal order is premised on the presence of an effectively normal situation in which the imminent validity of any particular legal norm becomes possible. But as Schmitt (1985: 13) notes, it is the 'sovereign who definitely decides whether this normal situation actually exists' and thus whether the enactment of a particular legal norm is necessary. In contradistinction to theorists such as Max Weber who defined sovereignty in relation to holding the monopoly on the legitimate use of violence, Schmitt argued that sovereignty resides in the monopoly to decide when the law – as a legal order – holds and when it does not. Because of the centrality of the decision (that a legal order holds) to his understanding of sovereignty, Schmitt critiqued liberal conceptions of sovereignty and the law as primarily second order sets of procedural rules. He therefore claimed that 'the legal prescription, as the norm of decision, only designates how decisions should be made, not who decides' (Schmitt 1985: 32–33). Moreover, he asserted that 'the legal idea cannot translate itself independently' (Schmitt 1985: 31). Thus, in seeking answers to the questions of if law, which law, and how law applies, Schmitt (1985: 34) argued that 'what matters for the reality of legal life is who decides'.

Schmitt's profane reading of the law reveals the legal realm and sovereign power to be dialectically constitutive. This is potentially instructive to understanding what enables the use of violence, including his functional assessment of the exception, its importance to the constitution of legal orders, and their limits. Schmitt's (1996: 66) critique extended further still to the rule of law which he argued was nothing more than 'the legitimation of a specific *status quo,* the preservation of which interests particularly those whose political power or economic advantage would stabilise itself in this law'.

Schmitt's understanding of exceptionalism – which serves to critique both legal positivism and liberal understandings of sovereignty – produces a particular form of politics. As Huysmans (2008: 169) argues, it is a politics

> structured at the interstice of law and executive government which is structured by the spectral question of when the necessity for legal transgression and political decision in democracies flips into the constitution of dictatorship.

It is thus a politics of a contested constitutionalism where law and sovereignty engage with one another dialectically. And this dialectic is premised on an original position in which there are clear lines of division between the law and sovereign power. As Huysmans (2008: 170) suggests, Schmitt develops a schematic in which 'the authentic nature of the political act is a decision that cannot be constrained by any normative foundations'.

The particularities of how the exception functions and its political consequences have been revived more recently by Giorgio Agamben. While Agamben's starting premises are very much indebted to Schmitt, his central problematique regarding the exception is different. Rather than see the ongoing dialectic between the law and sovereign power as the source of the exception, he diagnoses the problem as one in which the law and sovereignty have collapsed into one another. For Agamben (1998: 15), the paradox of sovereignty (i.e., law making power) is that the sovereign is '…at the same time, outside and inside the juridical order'. This means that sovereign power both upholds the law as well as determines exceptions through the law, instances in which sovereign action is not enveloped in the normal fabric of the legal order. In other words, 'sovereign power always retains an arbitrary, unmediated capacity to impose rule' (Huysmans 2008: 172).

Provocatively for Agamben (1998: 20), '…in our age, the state of exception comes more and more to the foreground as the fundamental political structure and ultimately begins to become the rule'. Such a claim requires that he establish the contours of the relationship between rule and exception. Agamben (1998: 18) argues that it is unhelpful to conceptualise them as a pairing in which the presence of one is independent of the other. Rather, the relation of exception in which something '…is included solely through its exclusion' involves a paradoxical form of mutual constitution:

> The exception does not subtract itself from the rule; rather the rule, suspending itself, gives rise to the exception and, maintaining itself in relation to the exception, first constitutes itself as a rule.
>
> (Agamben 1998: 18)

The exception is therefore central to the legitimacy and validity of the politico-juridical order by establishing the space in which it can be said that the order holds (Agamben 1998: 19). More important, every rule that forbids something has within it a presupposed exception that can bring about the rule's transgression (e.g., laws against homicide do not always apply in the killing of a human being by the state or its agents).

But the exception is not necessarily a normative suspension of the type outlined by Schmitt – or feared by liberal theorists – where the law is said to have disappeared (Huysmans 2006a). In other words, as conceptualised by Agamben, the exception does not suggest that sovereign power places an action, individual, or territory outside of the law so that the legal regime no longer applies. It is therefore not a suspension of the system itself but rather a tactical recognition of the limits of the scope of the law. For Agamben (1998: 28), the exception implies a form of abandonment in which sovereign decisions are rendered so that one '…is exposed and threatened on the threshold in which life and law, outside and inside, become indistinguishable'. While there is nothing outside of the law in Agamben's topography, the law remains predicated on the sovereign power of indistinction, the privilege to permit violence to pass over into law and the law to pass over into violence as required (Agamben 1998: 32). It is the tactical mobilisation of juridical ambiguity *par excellence*. Thus, the invocation of a state exception is less a 'spatiotemporal suspension' of normative values than the instantiation of a logic in which the exception and the rule, citizen and non-citizen, normal and abnormal, friend and enemy collapse, contributing to a form of territoriality without borders and a biopolitics without limits (Agamben 1998: 37).

Agamben (1998: 51–53) therefore claims that the exception in part operates through the structure of the 'sovereign ban' in which the law that is in force does not 'signify'. This means that under the exception, the law becomes ambivalent in its embracement of procedural norms and symbolic values. As a result:

> One of the paradoxes of a state of exception lies in the fact that in the state of exception, it is impossible to distinguish transgression of the law from the execution of the law, such that what violates a rule and what conforms to it coincide without any remainder.
>
> (Agamben 1998: 57)

To reiterate then, the criteria of the exception for Schmitt is a suspension of the law such that unlimited state authority – and an order supported by this authority – remain. The law and sovereignty are thus engaged in a dialectical relationship. The pragmatic pay-off for the Schmittian sovereign is that truly authentic political acts need not be constrained by normative foundations. In contrast, Agamben dismisses the Schmittian dialectic, the clear distinctions between law and sovereignty, and the importance of the suspension of the law for exceptionalism. Agamben's claim is that the exception produces a logic of violence and associated forms of territoriality that are both enabling of exceptional practices but also of the normal sovereign-juridical order. The political question then becomes how sovereign '...power that has crossed the [normative] threshold and thus lost its legality can nevertheless be legitimate' (Huysmans 2008: 172). More importantly, within the politics of the exception, Agamben argues that debates regarding normative suspensions (of the law) serve profoundly ideological ends. They reinforce the liberal notion that the law serves to constrain practices of power as well as realist notions that power must transcend the law within a geopolitical context so that '...the practices that are deployed are radically detached from any legal framework' (Huysmans 2008: 173).

The discussion above thus provides two distinct sets of criteria for identifying the politics of the exception. These will serve as analytics in positioning the rationales offered by Israel and the United States in relation to the broader legal problematisation of targeted killing; however, the targeted killing assemblage itself is shaped by the technical rationalities of the law rather than the suspension of the legal order. Thus interpretations and judgements about relevance, the concepts that help to define the relevant law, how law should be applied, and for what ends it must be so applied are central to the legal problematisation.

Most important here is that the law itself, and the way in which targeted killing is problematised as a legal concern, are reflections of broader relations of power that benefit those conducting counter-insurgency over insurgents. The law as a tool of political power, and political power as a tool of law, structure both the problematisations and their apparent reconciliations. At the same time, portrayals of law as a technology devoid of political substance attempt to mask the inherent politics of the law, the way in which political considerations shape legal reasoning, and our understandings of what constitutes the very conduct of war. It is therefore understandable that in attempting to protect this system of power relations and practices, contemporary governance relies on one of its most technical aspects of legitimacy – law – in order to shape the problematisation of what are the correct uses of force in counter-insurgency?

Differentiating assassination and targeted killing in legal problematisations

How then might the law normalise targeted killing and create space for its commissioning that does not rely on a politics of the exception? In mapping out how targeted killing becomes normalised within legal problematisations, the first step is a move to differentiate the practice from similar forms of violence, in terms of who is commissioning it, its spatio-temporal location, and how it is undertaken. Through this process, as discussed in the introductory chapter, targeted killing is differentiated from assassination. Assassination is generally conceived as a form of selective murder of an individual undertaken for political purposes (Harder 2002: 5). It is assumed that the practice itself is being undertaken by a state or its agents. With states continuing to hold the legal monopoly over the means of violence in international law, the commission of any killing or violence organised by a non-state actor is usually considered illegitimate – if not illegal – from first principles in both national and international law.

For the purposes of legal reasoning, a two-staged distinction is made regarding state-initiated individualised killing. First, it is important for the problematique whether an action takes place during a time of peace or during a time of war. If it takes place during a time of peace, it is not an assassination but some other act of killing that may or may not be prohibited under domestic or international law. Within existing legal frameworks, an assassination takes place during a time of war – as will be shown later, the distinction between these two conditions can be contentious. Emanuel Gross (2001: 241–242) argues that an assassination can be said to have taken place if two conditions are met: the aim of the action is to kill (a) particular person, and second, that the killing uses treacherous fighting tactics. If one of these conditions is not met, there is no assassination. The claim that assassination involves both the selective targeting of a specific individual for killing and the use of treachery and/or perfidy is a central aspect of the legal problematisation (Bazan 2002; David 2003; Guiora 2004; Havens et al. 1970). There is general agreement that what constitutes treachery and perfidy are those conditions laid out in the *Hague Convention* (1907) and the *Geneva Convention.* According to Article 23 of the *Hague Convention* (1907) it is forbidden to use poison, kill or wound 'treacherously', kill or wound an enemy who has surrendered, give no quarter, inflict unnecessary suffering, make improper use of flags, uniforms, or insignias in order to deceive, to recklessly destroy property, and/or to ignore the rights of hostile nationals.

Unlike the other clauses listed that clearly identify prohibited behaviours such as the use of poison, most analysts agree that treachery – as outlined in article 23(b) – is vague. However, as Harder (2002: 7) argues, treachery has more commonly come to be associated with a breach of trust when '...a victim has an affirmative reason to trust the assailant', or, as Zengel (1991: 622) states: '...treachery itself is understood as a breach of a duty of good faith towards the victim'. Thus, Ashkouri (2001: 167) notes, 'while the term "treacherous" has not been defined...it is not regarded as prohibiting operations that depend on the element of surprise, such as a commando raid or other form of attack behind enemy lines'. Article 37 of the *Geneva Convention* deals specifically with perfidy. It is perceived as providing more clarity with respect to what is permissible and impermissible. Perfidious actions are said to include: feigning the intent to negotiate, feigning wounds or illness, and feigning non-combatant or protective status. However, Article 37 makes it clear that 'ruses' – actions intended to mislead the enemy such as 'camouflage, decoys, mock operations, and misinformation' – are permissible. Thus, it is not the selective targeting of an individual that sufficiently constitutes assassination in its legal meaning. Rather, treachery and/or perfidy are the defining features of the legal distinction during a time of war. And it is this legal linking to treacherous and perfidious practices that binds to longer-standing representations that have coded assassination as feminine, dishonourable, and a weapon of the weak.[4] Moreover, the referent objects for protection under these laws are formally recognised soldiers and military-political elites as opposed to non-combatants. At the same time, what exactly counts as treachery is more open to interpretation than perfidy which has a more clearly defined legal understanding. As such, even legal definitions of assassination are more tenuous than usually assumed in the literature because they rest on which conventions are taken as relevant, how these are interpreted – especially notions of treachery – and who is doing the interpreting and for what purposes. There is also a clear privileging of particular forms of violence committed by some actors – read the armed forces of recognised states – as more legitimate than others. Yet, the way in which assassination is defined legally creates space for other individualised forms of killing.

Targeted killing

In contrast to assassination, targeted or selective killings are potentially more ambiguous with respect to their legal status. The crux of contention is summarised by David Kretzmer (2005: 174) who argues that:

> the disparity in the attitudes taken towards 'targeted killings' reveals a fundamental disagreement not only regarding their morality or legality, but also on the issue of the legal regime by which that legality should be judged. The states involved claim that such killings are legitimate means of fighting the 'war on terror', whose legality must be judged on the basis of the laws of armed conflict; those who label these killings 'extra-judicial executions' rely on a law-enforcement model of legality, which rests primarily, though not exclusively, on standards of international human rights law.

Similarly, George P. Fletcher (2006: 897) argues that determining the legitimacy of targeted killing under international law rests on how the threat/situation is defined (i.e., acts of crime or acts of war). Therefore, how terrorism is understood and defined legally is not only central to the legal problematisation of targeted killing, but also central to the practices of counter-terrorism.

Within legal discourses, terrorism has been discursively positioned as blurring the gap between crime and war (e.g., Crona and Richardson 1996; Brooks 2004). Therefore terrorism occupies a legal ground that fosters ambiguity, allowing states to select amongst provisions offered under criminal law and the laws of war – particularly in terms of widening the scope of actions that can be taken against insurgent groups. This does not involve ignoring the law but rather problematising counter-insurgency as an exercise in risk elimination that requires a flexible response. Terrorism and related insurgencies are understood to be transnational, operating in various legally defined situations – international armed conflicts, non-international armed conflicts, and non-conflicts – and taking *ad hoc* forms that do not adhere to orthodox (para)-military organisational structures. Thus, terrorism and insurgency, given the legal complexities they raise, are presented as problems to which existing legal frameworks provide only piecemeal advisement in any given situation on matters of permissibility.

Mapping targeted killing as a legal problem begins with defining the environment within which targeted killing is being deployed: armed conflict or peace. Specifically, if targeted killing takes place within a context where a recognised international armed conflict is absent then the common legal argument is that it would be lawful only if commissioned to prevent an imminent attack orchestrated by an individual or group located within one's sovereign jurisdiction that could not be thwarted by other means. If the targeted killing takes place within a context understood to constitute an international armed conflict, it is argued that it may be legal so long as only combatants are being selectively targeted and other criteria under the laws of war are met including military necessity and proportionality. Yet the divide between war and peace is not an easy distinction to make within existing legal frameworks. In international law, as Yoram Dinstein (2005: 4) notes, '...there is no binding definition of war stamped with the *imprimatur* of a multilateral treaty in place'. As a baseline for discussion, academic scholarship has often relied on L.F.L Oppenheim's classic definition of war: 'contention between two or more states through their armed forces, for the purpose of overpowering each other and imposing such conditions of peace as the victor pleases' subject to the recognition that contemporary war is now often conducted as an intra-state exercise (quoted in Dinstein 2005: 5). Such a formulation seems remote from contemporary battle-spaces. In response – as will be shown below – this disjuncture has given states such as Israel and the United States a great amount of legal leeway for counter-insurgency in general, and targeted killing in particular.

Targeted killing in war: *Jus ad bellum* problematisations

The argument that the legality of targeted killing is best covered under international humanitarian law – also known as the laws of war – makes its commission subject to *jus ad bellum* arguments about when a state can use force and *jus in bello* arguments about what a state may do when conducting hostilities (Lotrionte 2003: 78). *Jus ad bellum* arguments assume the same basic principles that combine to constitute a field of problematisation. These can be described as follows: a state – if it has suffered from an armed attack – is entitled to defend itself forcibly so long as the measures used are militarily necessary, timely (i.e., are temporally relate to participation in hostilities), proportionate, discriminate, minimise suffering and taken against those responsible – to some degree – for the attack (Guiora 2004: 324; Falk 2014: 299).

These principles regarding when a state can use force are then applied to the use of targeted killing in legal arguments. Given its use in counter-insurgency, the concern that immediately emerges is the issue of pre-emption or what is sometimes referred to as 'anticipatory self-defence': is it legally permissible to target the leadership or members of an organisation – state or non-state – who you believe may be planning to attack? Lucy Martinez (2003: 145) argues that the doctrine of anticipatory self-defence was widely accepted prior to 1945 as a part of customary international law. The customary threshold was that laid out in the Caroline Doctrine which arose after an American ship was attacked by British forces in 1832. The requirements rest with the initiator of a pre-emptive strike to demonstrate that a threat was imminent, that the anticipatory response was designed as a protection against the threat itself – as opposed to the achievement of other geostrategic goals – that the response was proportional to the threat posed, and that the use of force was the last resort after all other options had been exhausted (Martinez 2003: 129–30).

With the formation of the United Nations in 1945 and the passage of the UN Charter into formal international law, the legal status of pre-emption and anticipatory self-defence became highly contested (Martinez 2003: 145). Legal consensus rests on the opinion that the key provisions within the UN Charter to which any use of anticipatory force must be measured are Article 2(4) and Article 51. These demand that states refrain from the threat or use of force subject to retaining an inherent right to individual or collective self-defence under armed attack. Yet, there is no agreement about whether anticipatory self-defence survived legally, either alongside or under Article 51 of the UN Charter. This has raised questions about its existence, meaning, and scope (Martinez 2003: 145).

Those who argue that anticipatory self-defence is still allowed under the UN Charter tend to rely on arguments outside of legal reasoning. There are three inter-related points with respect to risk elimination that follow from one another. The first is that advances in modern war-fighting technology have greatly increased the destructive capability of any attack, making the stakes of waiting to see if a potential adversary actually engages in aggression unacceptable. The second argument is that not only do the threats posed by insurgency take advantage of these technological developments but that insurgents themselves, who are not a party to these agreements, will feel no obligation to be bound by any type of international law – customary or formal – prohibiting the use of force or engaging in aggression. The third rests on the political risks of not pre-empting an attack in terms of allowing attackers to plan further attacks with impunity.

Both the Bush and Obama Administrations pushed this logic of risk elimination to its furthest extreme by arguing that imminence need not necessarily be a factor for a pre-emptive strike: capability could serve as a sufficient condition for anticipatory self-defence. Therefore, an appeal to logic and common sense is presented; supporters of the doctrine of anticipatory self-defence say that its prohibition would leave states unable to proactively protect themselves from attack. Instead, they would need to wait to react to an attack while suffering all of the costs – in human lives and infrastructure – that waiting could entail. Thus, the argument is that not recognising pre-emption as legally permissible would establish a higher burden on states at the potential receiving end of attack than their aggressors (Martinez 2003: 158–159).

Guiora (2004: 324) goes further with respect to targeted killing by positioning it as a specific form of anticipatory self-defence – or what he calls 'active self-defence' that is more restrained than traditional war fighting tactics, thereby helping to make armed conflict more 'humane'. He argues that targeted killing, '...if properly executed, not only enables the State to more effectively protect itself within a legal context but also leads to minimising the loss of innocent civilians caught between the terrorists (who regularly violate international law by using civilians as human shields) and the State' (Guiora 2004: 324). In this way, targeted killing as a form of anticipatory self-defence is argued to mitigate unnecessary suffering (see also Strawser [2013] on the moral imperative to use drones in targeted killing). These formulations of the legality of anticipatory self-defence and how it is provided for under international law for the purposes of conducting effective counter-terrorism operations is not universally shared.

Arguments against the legality of anticipatory self-defence take myriad forms. Some believe that the concept itself is inconsistent with the prohibition on the use of force outlined in Article 2(4) of the UN charter which is meant to provide stability to the global system (Martinez 2003: 158–159). From this perspective, pre-emption creates a political environment in which the probability that state actors will resort to force increases, an outcome said to be at odds with the intentions of the UN Charter. In other words, under a regime of anticipatory self-defence, risk is not reduced, it increases exponentially. Moreover, anticipatory self-defence rests on the assumption that situations where an imminent attack is forthcoming are easily verifiable and not subject to misperception, bias, prejudice, or manipulation. Thus, as a counter to the arguments made by proponents above regarding the common sense basis for pre-emption, opponents make a contradictory appeal to common sense: given the destructive power of contemporary weaponry, it seems highly destabilising to the international system to allow states to pre-emptively strike one another based on a perception of threat. Anticipatory self-defence is a double-edged sword; if it is legal under international law, it becomes a potential tool for all states, a condition often overlooked by its most vociferous Western proponents (Martinez 2003: 180).

Practical objections to the legality of anticipatory self-defence extend beyond the ambiguity of imminence. These are often linked to understandings of contemporary geopolitical conditions. For example, it is argued that proportionality – at the best of times highly contentious in terms of how it is perceived – becomes even more difficult to judge in pre-emptive counter-insurgency operations. Martinez (2003: 175) notes that guidelines to proportionality can be found in the Additional Protocol to the 1977 *Geneva Conventions* under articles 57 and 58. She continues that proportionality demands that in any military operation, all necessary precautions should be taken 'to spare the civilian population, civilians, and civilian objects' (Martinez 2003: 175). These could include providing advanced warning of an attack, removing civilians from the vicinity of military objectives, and avoiding the placement of military objectives near densely populated areas. Furthermore, it is expected that no military operation shall cause 'incidental loss of civilian life, injury to civilians, damage to civilian objects, or any combination thereof, which would be excessive in relation to the concrete and direct military advantage anticipated'.[5] But the calculations here with respect to military advantage and civilian losses are not clear cut.

Martinez (2003: 176) argues that respecting these articles requires that the use of force only targets the terrorist or insurgent group and must be proportionate to the anticipated act. For others, a strike is proportionate based on its relation to the future threat posed, as defined by past practices of the individual or group, their motives, the current context, and preparatory actions for attack that have been undertaken (Downes 2004: 288–289). Yet, as these involve calculations of risk, these are not necessarily objectively determinable qualifications. Falk (2014) for example documents the variations in calculations of proportionality within the Israeli context and the impossibility of any *a priori* objective standard. Beyond the case of Israel, it is said that as perceptions of the destructive and adaptive capabilities of terrorist and insurgent groups escalate, those who determine the baseline for proportionality must wrestle with the question of how to calculate the incalculable. Therefore, the accountants of proportionality have great scope in their evaluation of the threat posed – or military advantage gained. The already low reliability of intelligence on the potential scope and consequences of a possible attack by a state actor is subject to even greater levels of indeterminacy when dealing with a perceived threat from insurgency groups. Insurgent movements are usually not subject to many forms of independent capability monitoring like weapons inspections, self-reporting, or the submission of documentation for technologies which are in their possession. This can lead to an exaggeration in the risks posed.

In addition, given that international law is premised on the maintenance of a global order defined by sovereign states with clear boundaries of territorial authority, anticipatory self-defence raises sticky issues with regards to both sovereignty and territoriality. Martinez (2003: 179) formulates the problem thus:

> …any unilateral use of force [by State A] against State B on the basis of anticipatory self-defence against an imminent terrorist attack must, at the least, be predicted upon a nexus between State B and the terrorist organization. The question of how substantial that nexus must be is extremely complicated and contested.

Thus, there again is considerable scope for manoeuvre under the current legal regime governing violence.

As discussed above, anticipatory self-defence by its very definition is considered to be a proactive, forward looking, and pre-emptive means of risk elimination. It involves the use of force to commit a first strike in anticipation of an imminent attack by another party. Yet many positions in support of targeted killing rely on moral arguments about retribution claiming that it best matches punishment to the crime of killing others through the commission of violent acts (e.g., David 2003; Statman 2004; Patterson and Casale 2005). However, retribution does not fit into the anticipatory self-defence framework or its principles of legitimacy. A predominant legal view is that if retribution is given as a reason justifying a specific act of targeted killing, then the act itself is not an instance of anticipatory self-defence under current legal definitions. This means that either the perpetrator of the targeted killing is involved in a standard armed conflict or that the killing should fall under the domain of international human rights law. If the claim is that the act has occurred during an armed conflict, the standard laws of war apply – including protections for enemy combatants as well as limitations on the means that may be used to kill them. Thus, the military leadership of a force may be selectively targeted for elimination but the laws of war prohibit the named targeting of individual soldiers.

If the violence is said to have occurred outside of an armed conflict, then the presence of retribution as a motive can be taken to legally imply that an extra-judicial form of killing has taken place. If this act is not a direct violation of domestic legal statutes, it is commonly understood to be in direct violation of existing international human rights law. Extra-judicial execution could be found in conflict with Article 3 of the *UN Declaration on Human Rights* which states that 'every person has the right to life, liberty, and security of person'. It could also contravene the first principle of the *1989 UN Principles on the Effective Prevention and Investigation of Extra-legal, Arbitrary and Summary Executions* which argues that 'exceptional circumstances, including a state of war or threat of war, internal political instability or any other public emergency may not be invoked as a justification of such executions'.

In summary, *jus ad bellum* problematisations are strongly influenced by a more general imperative to reduce risks to state security, both from physical attack but also in terms of legal liabilities for sovereign state actors when using force against insurgents. The doctrine of anticipatory self-defence, a high-profile tool of contemporary counter-insurgency, is therefore central to the ways in which arguments in favour of targeted killing are made and provides considerable scope for the practice. Moreover, it presents clear limits, including the illegitimacy of retributive justifications that can be deployed, for the purposes of stabilising targeted killing as a disinterested risk management practice. But what about during times when hostilities have already commenced?

Jus in bello *problematisations*

Unlike *jus ad bellum* considerations, what a state may do while conducting hostilities is more concrete in terms of the mechanistic processes that must be followed and the calculations specific to these processes that must be made. As noted above, there are very clear rules about the intentional killing of specific individuals during a time of conflict: there can be no treachery or perfidy, any use of force must demonstrate military necessity, and any use of force that might kill or injure civilians must abide by the principle of proportionality. A violation of any one of these requirements and the action could be declared to be either an act of assassination and/or a war crime in a legal setting. However, there are additional elements that form a part of the *jus in bello* legal problematisation. These involve technological considerations, distinctions made between the military leadership and 'common' soldiers, how (or whether) to distinguish between regular forces and 'irregular forces' like insurgency groups or terrorist organisations, and proportionality.

Ward Thomas (2000) has noted that in a declared war, there is no legal prohibition against selectively targeting military leadership though there had been, until the late twentieth century, a norm against it. Thus, there is a general legal consensus in a conflict that one can target the leaders of a rival military organisation for elimination. The qualifying conditions – as noted above – are that the means do not involve a breach of duty to the intended victim or a betrayal of the victim's confidence with respect to protections that should be accorded.

In part, the renewed interest in assassination and targeted killing as legal problems stems from technological developments including improvements in satellite surveillance, the increased capability to intercept signals communications, and remotely piloted aircraft (RPAs) that are argued to have made many of the potential treacherous or perfidious aspects of assassination – like double-agents, informants, or other forms of betrayal to increase an adversary's vulnerability to attack – moot. Thus, new forms of technological sophistication have shifted *jus in bello* problematisations away from treachery and perfidy as objects of investigation towards the broader legal implications of targeted killing, the measurement of proportionality, as well as the issues of sovereignty and territoriality.

While the targeted killing of enemy military leaders is argued to have a substantial foundation for its claims to legality, there is less certainty with respect to the targeting of named individual combatants. In other words, although individual soldiers may be targeted and killed as a part of normal operations to achieve specific military objectives, there is disagreement over whether a specific combatant can be named and then pursued for elimination as a military objective in and of itself. Customary law, including military handbooks on ethics, would seem to prohibit the practice, in part because it could be seen as unduly punishing soldiers for playing their roles in combat, thereby placing it at odds with other protections provided to combatants under the Geneva Convention. But in counter- insurgency it can be argued that there are tactical and strategic advantages to targeting an individual because of his/her specific attributes (e.g., bomb-making expertise) or their strategic position within a network. Thus, two ways have been used to abide by the letter of the law. The first is to make a claim that a named individual occupies a leadership position within an organisation. The second has been the shift to signature strikes (i.e., attacks aimed at individuals or groups demonstrating an 'insurgent pattern of life') which then are a part of normal combat operations seeking to achieve specific military objectives.

Whether practices of targeted killing can be legally condoned is also argued to rest on the status of individual participants in armed conflicts which may involve questions of temporality in terms of the relationship between acts of organised violence and the combatant status of those who may have contributed to them. The current focus is on determining the status of 'terrorists', as well as longer-standing questions regarding 'guerrilla' fighters or 'insurgents'. Gross (2001: 206) thus formulates the question as such: 'is it proper to regard terrorists as combatants, and thereby grant the terrorists the protection due to combatants, *a fortiori*; and is it improper to regard terrorists as civilians who are not combatants, and grant them even more extensive rights?'

It is permissible to kill combatants in an armed conflict, keeping in mind the restrictions noted above and the differences in opinion on who may be selectively targeted for elimination. However, who should be granted combatant status is ambiguous in current legal frameworks. This has become a pressing issue in contemporary conflicts. Should 'terrorists' or insurgents be considered combatants, and if so, for how long: while committing specific acts of hostility or permanently until all hostilities cease? If they are treated as combatants, along with Geneva Convention protections, there would still be the potential to provide a legal foundation that could be used to advance arguments justifying the option to selectively kill them. However, as Kretzmer (2005: 199) notes, it is an unresolved question as to whether having membership in a group labelled as a terrorist or insurgency organisation, necessarily means that one is taking a direct part in hostilities, a key aspect of the combatant designation. Leaving the difficulty noted above of legally defining hostilities, how would one define taking 'direct part'? Is it during the commission of a particular act or does participating in one act permanently brand someone as a combatant? These points are not merely hypothetical: many targeted groups have mixed political and military structures, much like the state itself. Kretzmer (2005: 203) opines, assuming that all definitional caveats could be resolved, that only actors who are members of the military wings of these organisations who are taking direct part in hostilities – or planning hostile activities – should be understood as legitimate targets. Thus, in this formulation, targeting an organisation's political leadership would be illegal as would be targeting former members.

Status also raises another set of questions that creates a space for targeted killing. If 'terrorists' and 'insurgents' are to be understood as non-combatants: should they be subject to international human rights laws and therefore approached under use of force doctrines applicable to criminals? Or, should terrorists and insurgents fall between the two distinctions, leaving them liable to facing measures under both types of law? As Michael L. Gross (2006: 324) argues:

> From any informed perspective, they [terrorists and insurgents] are not, as some human rights advocates seem to suggest, civilians who occasionally and only marginally contribute to armed struggle. On the contrary, they maintain their hostile status off the battlefield as they prepare for battle, lay plans, tend to their weapons and maintain their fighting capability. At the same time, there is good cause to suspect that terrorists are guilty of war crimes and criminal activity.

Although, the formal legal status of terrorists and insurgents remains contested, as will be shown, legal determinations are made on an *ad hoc* basis by states to demonstrate that their counter-insurgency operations are legally permissible, rather than declaring that laws are being suspended to enable the practice of targeted killing.

Proportionality

As discussed above with respect to anticipatory self-defence, the principle of proportionality recognises that in any military operation there is the very real possibility – if not certainty – of inflicting civilian causalities. As such, the first consideration for *jus in bello* determinations is whether a particular goal can be achieved without recourse to attack. In an armed conflict this may be impossible. If so, the claim is that the condition of proportionality is meant to minimise civilian causalities, regardless of the significance of the military objective. Proportionality dictates that civilians may not be intentionally targeted and that civilians should be given warning – if feasible – to remove themselves from the vicinity of 'legitimate' military targets. The principle of proportionality is also widely interpreted as requiring that any civilian casualties and/or other forms of damage must be in proportion to the importance of the military objective. The more important the military target, the more leeway for 'collateral damage'. Thus proportionality is not determined by the weapons system used (i.e., a particular system is not inherently more proportional than another); rather proportionality is determined by the probable loss of civilian life in relation to the importance of the military objective (e.g., Braun and Brunstetter 2013).

It is argued that judgements of proportionality and the necessary measures to abide by the principle should take place in anticipation of the commission of any military strike. They may also be revisited in retrospect. However, many argue that in revisiting calculations of proportionality, one must only pay heed to the information – or misinformation – reasonably available to military commanders at the time of ordering the attack. As Mary Ellen O'Connell (2002: 35–36) suggests:

> responsibility for mistake is a debated point in international law. Some writers suggest that if the state taking enforcement measures was mistaken regarding the existence of a wrong or the gravity of the wrong, its response based on that mistake should be excused as long as the state acted in good faith. State practice confirms this.

It is here where the ability of insurgents to blend into the civilian population – by not wearing uniforms or bearing arms openly at all times – which has the potential to be a difficult problem, can be elided. The difficulty or inability to distinguish a combatant from a non-combatant in a given theatre of conflict has become a *de facto* license to excuse making finer grain distinctions in practice. For example, this cessation underpins the American practice of signature strikes and the way in which any males of military age who are killed in such operations are automatically designated as insurgents.

Proportionality as a legal concept though assumes that the kinds of trade-offs upon which the concept relies are actually empirically verifiable and that the potential exists to measure them 'objectively'. Moreover, it gives latitude to take the purported rationales, calculations, and understandings of (victorious) military commanders at face value. As such, the vision of legality – or even justice – underpinning the problematisation of proportionality is procedural (i.e., were reasonable calculations made prior to the attack?) as opposed to substantive. And such calculations not only work within the law, but are a requirement of it.

Sovereignty and territoriality

As in *jus ad bellum* problematisations, sovereignty and territory play a significant role in shaping *jus in bello* problematisations of targeted killing. Primarily, this is framed around determining the legality of conducting targeted killings outside of a state's sovereign territory, an act usually interpreted within legal reasoning as a form of aggression, if not an act of war. As Downes (2004: 292 quoting Paust 2002) has noted though, some argue that these attacks do not constitute a traditional use of force against a state in the sense that 'territorial boundaries and regimes are not changed' and therefore should not be interpreted negatively.

Regardless, as a matter of practice, nearly all targeted killings take place outside of the sovereign jurisdiction of the state committing the act, with a significant number conducted in third party territories not formally engaged in hostilities. Furthermore, sovereign territory and sovereign control are not synonymous. From Jewish settlements on the West Bank to Pakistani provinces along the Af-Pak border to the interior of the Democratic Republic of the Congo, a territory may be officially recognised in international law as being under the sovereign authority of a specific state with that state actually exercising very little *de facto* control. But a key question is whether the legal problematisation is responding to the practices of targeted killing or in fact helping to determine the extra-territorial aspect of contemporary targeted killings.

Given the common understanding within diplomatic culture that a state committing an act of political violence in another state's sovereign territory is a significant breach of peace subject to sanction under the UN Charter, how has it become possible for many targeted killings to take place outside of the recognised war zones, let alone the sovereign territories of states engaged in hostilities? In part, it may be because it is more difficult to legally justify targeted killing within one's own sovereign jurisdiction (as will be seen below). Pragmatically, the burden of proof to demonstrate that other means could not have neutralised the threat in question – such as an arrest – become more onerous. Legally, there is an assumption that the norms of international human rights law – often cornerstones of liberal rule – will hold within the domestic sphere. Moreover, for countries that have previously committed targeted killings such as the United States, Israel, France, South Africa, and the United Kingdom, there may also be a sense that to conduct these types of operations domestically without a compelling narrative could have significant political repercussions.

But given the number of targeted killings taking place in third party states who are not formal parties to recognised armed conflicts such as Pakistan or Somalia, the *jus in bellum* legal problematisation does not emphasise the need for any formal procedures of acceptance of the targeted killing in their territory by these states. For example, in a targeted killing committed in 2002, Yemeni authorities made it clear that an operation conducted by the US military in Yemen was taken with their consent. In sharp contrast, Pakistani authorities are publicly critical of American targeted killings of suspected Taliban insurgents taking place on the Pakistani side of the Af-Pak border. Yet evidence suggests that the Pakistani government has at least given implicit permission for some of these operations to take place (Orr 2011: 730); whether consent has been coerced or incentivised is a different question. Either way, there are no established legal procedures for obtaining or demonstrating consent for such operations. Thus, legal concerns over sovereignty and territory are not being addressed within the legal realm despite their centrality to the legal problematisation of targeted killing. In this sense, the legal problematisation occludes a procedural omission that allows the fate of persons of interest to be horse-traded outside of any formal oversight or review established under international law. This particular aspect of the politics of elimination, despite an abundance of rules, regulations, and legal principles otherwise, is thus one that falls outside of formal regulation. This could perhaps reflect a cultural aversion to public demonstrations in which human life is treated as negotiable capital. But it also enables a de facto expansion of a particular form of territoriality (i.e., a particular set of rationalities for governing territory) into the sovereign space of others.

Problematising targeted killing under international human rights law

The preceding analysis has outlined the ways in which targeted killing has been problematised under international humanitarian law or what is often more commonly known as the laws of war. However, the claim that international humanitarian law is the appropriate body of law is itself highly contested. Some argue that any discussion over the legality of targeted killing must turn to the protections offered under international human rights law. In contrast to dominant narratives of the 'war on terror' that script its activity as military operations, there are those who argue that counter-insurgency is best seen as a type of policing action, making it subject to traditional limits on the exercise of arbitrary authority as well as strict rules on the use of force (see Banks [2013] for a recent exploration of these positions and their implications).

Kretzmer (2005: 176) provides an overview of the legal problematisation of targeted killing under international human rights law. The first question to be considered is, are there circumstances in which a targeted killing would not be regarded as a violation of the victim's right to life? The second question regards sovereignty and the scope of international human rights law: does it apply to actions undertaken by State A in the territory of State B over which State A has no jurisdiction or control? The third issue is whether international human rights law can apply to situations that are 'warlike' but not officially recognised as an armed conflict between an international insurgency movement and a specific state (Kretzmer 2005: 177).

The inherent right to life is supposed to be protected even in times of emergency. However, Kretzmer (2005: 177) notes that non-derogation is not legally the same as absolute protection. The *International Covenant on Civil and Political Rights*, for example, prohibits 'the arbitrary deprivation of life' making the determination of a violation tied to judgements of what can be said to be arbitrary. The *European Convention for the Protection of Human Rights* (ECPHR) offers a different standard. Kretzmer (2005: 177) notes that Article 2(1) of the ECPHR states that no one should be intentionally deprived of their life save for circumstances in which it 'results from the use of force which is no more than absolutely necessary...in defence of any person from unlawful violence', in order to effect a lawful arrest or to prevent escape of a person lawfully detained, and/or in action lawfully taken for the purpose of quelling a riot or insurrection. With regards to targeted killing, the first special case is seen to offer the most plausible justification for its commission.

Kretzmer (2005: 178) argues though that there are further considerations that must be held to account. First, is the use of force absolutely necessary or are there other measures available that could protect people from unlawful violence? Second, if there are no other measures available other than the use of force, is it absolutely necessary to use lethal force or could non-lethal force achieve a similar intended result? These questions themselves are predicated on principles of due process that are central to human rights law and a staple of liberal rhetoric within the Anglo-American orbit. First, individuals are innocent until proven guilty. Second, persons suspected of engaging in or planning criminal activity are to be arrested, detained, and/or interrogated under the due process of law. Third, if there is credible evidence of involvement, those who are accused should receive a fair trial before a competent and independent court, and if convicted, be subject to a punishment provided by law.

As Kretzmer (2005: 178) argues, under this model, a state cannot simply eliminate someone that it believes may be about to commit an act of unlawful violence in order to prevent that violence from taking place. Prevention, in the first instance, should be achieved through apprehension (i.e. lawful arrest), and in the second, by subjecting suspects to the mechanisms of criminal law. This means that force cannot be applied '...unless it is clear that there was no feasible possibility of protecting the prospective victim by apprehending the suspected perpetrator' (Kretzmer 2005: 178). Even more hawkish analysts like Guiora (2004: 330) concede this point, noting that one of the elements differentiating targeted killing from extra-judicial execution is the impossibility of affecting an arrest. However, contemporary conflicts are not usually bound by the territory of a given sovereign state. Thus, Kretzmer (2005: 178) identifies a key weakness in international human rights law: its inability to provide protections in instances where a suspected perpetrator is not within the jurisdiction of the 'victim' state so that an arrest can take place. While there is always the possibility that State B might adhere to the request to apprehend, detain, and extradite an individual or group present in its sovereign territory that poses an imminent threat to State A, State B could be unwilling or even lacking the capability to do so. The question then is does a lack of willingness and capability reduce the force of the right to life in these instances?

The legal regime provides no clear answer. As Kretzmer (2005: 180) suggests, the *UN Convention on Civil and Political Rights* excludes any specific discussions of instances in which the right to life may be legitimately violated. In a decision made by the Human Rights Committee in assessing Israel in 2003, it was explicitly noted that before deadly force may be used, all measures to arrest a person suspected of being about to commit acts of violence must be deployed. Yet, if a suspect is planning attacks against a state and prevention through arrest is not possible, international human rights law is unclear as to whether this would constitute an arbitrary deprivation of life. Kretzmer (2005: 182) argues that in these situations, imminence becomes the key principle towards providing a credible legal justification for targeted killing. Rather than rely on arguments about 'the last window of opportunity to frustrate further terrorist attacks' that are often a staple justification in instances where an attack is not thought to be imminent, imminence makes the use of due processes methods problematic while potentially providing evidentiary grounds for killing a suspected terrorist or insurgent (i.e., the evidence is there to be seen) (Kretzmer 2005: 182). And thus, this focus on imminence, itself difficult to measure or verify, creates a potentially powerful means to legally displace the right to life within the contours of the law.

The way in which sovereignty and territoriality are inserted into legal problematisations of assassination and targeted killing is also said to raise issues of jurisdiction. Normally, all state parties to these agreements are obligated to adhere to their provisions to all those within its territory and subject to its jurisdiction. The problem is to determine whether restrictions on the arbitrary deprivation of life apply in cases where a state is not party to the relevant legal conventions or when persons targeted for killing are residing within a jurisdiction that is not a party to international human rights agreements like the *UN Convention on Civil and Political Rights*. Kretzmer (2005: 184) argues that the customary law and norms underpinning these agreements are intended to be universally binding. If not, a situation would be created where it would be lawful for states to kill persons in other sovereign jurisdictions outside of armed conflict. However, imminence provides a legitimating rationale for eliminating individuals though this legal permissibility might be contested.

National legal problematisations

Beyond laws – both statutory and customary – norms, and regulations that arise from within international legal frameworks, there are unique problematisations that are shaping legal understandings of assassination and targeted killing within specific national jurisdictions. Unsurprisingly, given their level of involvement in acts of assassination and/or targeted killing in counter-insurgency operations, the two states in which national laws contribute the most to legal problematisations are the United States and Israel. In Israel, issues surround the legal status of Palestine with respect to Israeli governance, whether Israel is engaged in an armed conflict or policing action with Palestine and the applicability of the Basic Law to instances of targeted killing. Although the Israeli Supreme Court initially refused to review the policy of targeted killing in 2002, it revised its opinion on its appropriateness as a venue, and ruled that targeted killing was consistent with the Basic Law in 2006. Within the legal context of the United States, debate initially focused on the meaning and power of the Executive Orders on assassination (11.905, 12.036, and 12.333) and if these place additional restrictions on targeted killing not covered under international law. More recently, legal controversy has shifted to the use of drone strikes and the targeting of American citizens residing abroad for elimination.

Israel

Within Israel, the legal problematisation of targeted killing has been framed by two issues. The first is the status of Palestine vis-à-vis Israel. Whether Palestine is to be interpreted as a separate sovereign state, as a quasi-independent state that is occupied by Israel, or another designation has important consequences. Similarly, competing interpretations of the rights guaranteed under Israel's Basic Law are also central to competing arguments over the legal compatibility of targeted killing with national jurisprudence.

The status of the Israel–Palestine dispute

The status of Palestine with regards to Israel is important according to Kretzmer (2005) because it signals the set of laws – international humanitarian or human rights – that might be appropriate to clarifying the legal status of Israeli targeted killing. Primarily, should attacks on Israeli territory and civilians orchestrated by Palestinian insurgents be defined as acts of war, acts of transnational terrorism, or acts of criminal homicide? The status of these events has not been tied into the specifics of the events themselves (e.g., where the attack took place, how many people were injured or killed, or what type of attack occurred). Rather, it was the legal status of the West Bank and Gaza with regards to Israel that has been considered to be the decisive legal factor.

Within traditional legal analytic modes, the first determination would be if the events satisfy common legal understandings of what constitutes an armed conflict. The second determination would be whether the Palestinian–Israeli conflict is constituted as an international or a non-international dynamic (i.e., a type of civil disturbance). At this point, an initial judgement would be made. If one were to judge that incidents between Israel and Palestine did not reach a required threshold to be considered an armed conflict, then targeted killing could be illegal except in the circumstances tied to imminence and last resort noted above under international human rights law.

If one believes that there is an armed conflict according to some legal definition, international humanitarian law only might apply. However, as

Kretzmer (2005: 205) argued, if Israel was legally considered to be an occupying power of Palestine, then Israel would be obligated to follow a law enforcement approach as outlined in international human rights law. If Palestine was legally considered to be an independent sovereign nation state, then however Israel wishes to respond to acts of violence committed, it would potentially need to be honouring the rights and protections offered to combatants under the relevant protocols of the Geneva Convention. Thus at the heart of the problematisation in the Israeli context is whether kinetic actions against Palestine constitute an international or non-international conflict? Over the past two decades, practice has shown that the art of government for the Israelis has been to pick and choose which legal status works best in a given situation. As such, a novel legal rationale has proven necessary for the Israeli government to target persons of interest but without giving their targets the associated privileges of being a combatant.[6] Thus the Israeli Defence Force takes the position that 'the setting of the conflict between Israel and the different Palestinian organization be referred to as a "state of armed conflict short of war"' (Falk 2014: 304). In theory, this provides the leeway to pick and choose provisions that offer the most scope for targeted killing from international humanitarian and international human rights law.

Constitutional rights and the Basic Law

Some analysts and legal professionals present arguments that frame the Israeli –Palestinian conflict as a civil disturbance best covered under international human rights law and specific constitutional rights found in the Israeli Basic Law. Arguments with respect to international human rights law have already been covered above so the focus in this section will be on the Basic Law. David (2003: 114) illustrates that Israel's Basic Law guarantees 'there shall be no violation of the life, body, or dignity of any person as such'. However, these rights can be suspended by any law 'fitting the values of the state of Israel designed for a proper purpose, and to an extent no greater than required or by such a law enacted with explicit authorisation within' (David 2003: 114–115). Moreover, in times of declared emergency, legal provisions exist to restrict or deny rights under the Basic Law. Therefore, Israeli constitutional law establishes its own limits and thus its suspension – while a sovereign prerogative – has a legally justifiable basis. But, there are rules to be followed. A 2002 recommendation by the Judge Advocate General of the Israel Defence Forces outlined the three essential conditions that must be satisfied before a targeted killing can take place. First, the Palestinian Authority must ignore appeals for arrests of persons of interest. Second, Israeli security forces must reach a conclusion through careful assessment that they cannot arrest the persons of interest themselves. Third, a targeting killing should only be done to prevent an imminent or future attack; retribution or revenge are not to be factored in. These conditions were supported in rulings given by the Israeli High Court of Justice (HCJ), particularly the 2006 decision which affirmed the legality of targeted killing.

To date, the 2006 decision is the only instance where a national Supreme Court has ruled on the legality of targeted killing and it is argued that it has '…become a widely accepted international reference, perhaps setting an international standard' (Falk 2014: 301). Expanding notions of justiciability in order to hear the case, the HCJ ruled that targeted killings could be carried out in the 'territories'. The decision though rested on the confirmation that guiding principles should be provided by the Laws of War. Specifically – and in line with IDF practices – the HCJ determined that six conditions must be satisfied prior to initiating acts of targeted killing. First, arresting the target is not feasible. Second, the target must be a combatant. Third, senior members of the Israeli cabinet must give their approval for all targeted killing operations. Fourth, civilian losses must be minimised. Fifth, operations can only take place in territories not under Israel's direct physical control. Sixth, the target must be identified as a future threat. In addition, the HCJ added the requirement of an independent investigation by the government following every instance of targeted killing to ensure that these key principles were adhered to and that required practices were undertaken.

Beyond establishing the legality of targeted killing and the thresholds that must be met to reach this status, the 2006 decision also attempted to clarify the status and rules applicable to civilians who take direct part in armed hostilities. The determination narrowed the privileges extended to combatants while widening what constitutes direct participation by civilians, reducing their scope for immunity. The consequence was the creation of a legal category of individuals who neither benefit from the privileges extended to combatants nor from the protections offered to civilians (Falk 2014: 301–308; see also Jones 2015a: 682).

In practice then, the law has been essential for enabling targeted killing in the Israeli context. Beyond legal statues and decisions, this has also meant a more direct role is being played by lawyers within the Israeli defence forces. Craig Jones (2015a: 677 italics in original) has demonstrated, military lawyers in Israel have moved from providing general advice about where and how military missions might take place to a role where '[they]…also *approved* and effectively *decided* the outcome of lethal operations'. As such lawyers have become central to 'the planning (deliberative) stage and the live (dynamic) stage' of targeted killing operations with advice given often leading to the expansion of permissible practice through the use of existing 'legal classifica-tion[s]' (Jones 2015a: 683). The expansion of permissible practice has been achieved by applying the abstract legal principles of international humanitarian law and constitutional law to concrete military engagements, including targeted killings. This process has been referred to in the legal literature as 'opera-tional law' (Jones 2015a: 690). As Jones (2015a: 690) suggests, 'operational law provides militaries with a vast discursive and practical apparatus to interpret, apply, and effectively remake international and domestic law in and through the sites, sights, and spaces of war.' As the case of Israel demonstrates, the law cannot only tolerate targeted killing, but can also become an essential aspect of its commissioning. Thus, the law has been proactively and effectively used by Israeli authorities to stabilise its targeted killing assemblage.

The United States

While it is taken for granted that the first Executive Order 12.333 drafted by Gerald Ford – and then subsequently revised and reaffirmed by Jimmy Carter and Ronald Reagan – sought to prevent assassination being used as a tool by US security services, there is still some contestation over the exact acts that the orders attempted to ban. Neither specific descriptions of prohibited activities nor a definition of assassination is presented in the Executive Order. Sean D. Murphy (2003: 363) believes that '...the wording of the Executive Order coupled with the context within which it was formulated and passed (i.e., under the Ford Administration) suggests that it was meant to prohibit the killing of government officials, not non-governmental persons...'. Similarly, Byman (2006) argues that from its inception, US administrations have interpreted the ban as not applying to use of military force to attack military commanders, even when these may be heads of state (e.g., Gaddafi and Hussein). Harder (2002: 2) believed that the ambiguity created a 'dangerous pitfall' in that it had 'the potential to artificially circumscribe US flexibility, or, at a minimum, create misplaced public enmity towards the military'.

Initially Ford's prohibition covered those employed by the US government only. This was extended by Jimmy Carter to include employees and those acting on their behalf, an extension that was confirmed by Ronald Reagan. Therefore, Bazan (2002: 1) outlines the restrictions enshrined within Executive Order 12.333 as follows. Section 2.11 stipulates that no one employed by the US government or acting on its behalf shall engage in or conspire to commit an assassination and Section 2.12 prohibits indirect participation, including requesting others at arm's length to undertake activities forbidden by the order. But this is still legally vague. Thus, part of the national legal problematisation around assassination and targeted killing for US security services and the military has been to offer a generously narrow interpretation of the order to provide a wide spectrum of tactical flexibility. For example, the US Army's Judge Advocate General determined in 1989 that

> the clandestine, low visibility or overt use of military force against legitimate targets in time of war, or against similar targets in time of peace where such individuals or groups pose an immediate threat to the United States citizens or the national security of the United States, as determined by competent authority, does not constitute assassination or conspiracy to engage in assassination, and would not be prohibited by the proscription in EO 12.333 or by international law.
>
> (Parks 1989)

Similarly, the legal opinion expressed by W.H. Parks (1989) is that 'acting consistent with the Charter of the United Nations, a decision by the President to employ clandestine, low visibility or overt military force would not...[be illegal] if US military forces were employed against the combatant forces of another nation, a guerrilla force, or a terrorist, or other organization whose actions posed a threat to the security of the United States'. In other words, so long as it presents a plausible narrative, it is possible for the US government to target whomever it wants for elimination outside its borders without fear of any domestic legal consequences under EO 12.333.

Thus, as Damrosch (1989: 800) argues, the legal weight of the Executive Order is minimal. It makes a public statement but can be 'countermanded by the president on his own authority'. Murphy (2003: 363) concurs with this assessment. He argues that while a standing US Executive Order might bar assassination, it could be amended by a presidential order or directive while Bazan (2002) and Gross (2001) noted, a ban could even be revoked by a president or Congress at any time. However, mechanisms for transparency would require that any change would have to be made public in the Federal Register 44 USC section 1505 (Bazan 2002: 2). While the ban has still not formally be removed, the passing of S.J. Res. 23 and H.J. Res. 64 on September 14, 2001 gave the president the authority to 'use all necessary and appropriate force against those nations, organizations, or persons he determines planned, authorized, committed, or aided the terrorist attacks...in order to prevent any future acts of international terrorism against the United States by such nations, organizations, or persons' (cited in Bazan 2002: 6). Moreover, under the spectral regulatory regime of 12.333, Koh (2010) has argued that, '...the use of lawful weapons systems – consistent with the applicable laws of war – for precision targeting of specific high-level belligerent leaders when acting in self-defense or during an armed conflict is not unlawful, and hence does not constitute "assassination"'.

In essence then, Executive Order 12.333 has created a domestic legal context that does not outlaw assassination and/or targeted killing. As Harder (2002: 2) illustrates, assassination was already illegal under international law. Cynically, it has been suggested that its issuance and resulting high profile status – despite its extremely limited legal power and ambiguity – served as a means to prevent the imposition of formal legislation that may have been more limiting in the types of tactics that could be used to exterminate persons of interest. In sum, rather than creating a political environment in which acts of assassination or targeted killing are prohibited, the order has constructed a dynamic in which it necessary to have received the approval of the executive branch of the US government before conducting these types of 'sensitive' operations. Its exercise therefore remains in the hand of the 'sovereign'. The specific role of the president in targeted killing is at the stage of 'authorising the target'. Thus while President Obama has been the final stage in the approval process for targets initially selected by the Joint Special Operations Command (JSOC) – a process that moves up the military chain of command and into the

Principles Committee – he does not necessarily approve each individual strike (Currier 2015).

Although authorisation for 'initialising' the sharp end of the kill chain may be left in the hands of the sovereign, this does not necessarily indicate the presence of a normative suspension as in the Schmittian conception of the exception (Currier 2015). With the frequency of targeted killing increasing under the Obama Administration, representatives of the US government have articulated defences of the practice that are explicitly linked to the existing legal order. Unlike Israel, which has constructed a legal justification under domestic law, the legal rationale and defence offered by the United States weighs more heavily on the scope for lethal force under international humanitarian law. At the same time, a desire is expressed to be able to work within a law enforcement paradigm, though for pragmatic rather than ethico-legal reasons. As Attorney General Eric Holder (2012) stated:

> It is preferable to capture suspected terrorists where feasible – among other reasons, so that we can gather valuable intelligence from them – but we must also recognize that there are instances where our government has the clear authority – and, I would argue, the responsibility – to defend the United States through the appropriate and lawful use of lethal force.

Thus, the claim is that in the war of global counter-insurgency, the United States has authority under international law to use force – including deadly force – to protect itself from attack. Their lack of conventional forces and their ability to blend into the civilian sphere is claimed to make operations against insurgents and the application of international law in these operations more difficult. However, it is also claimed that the application of international law is critical for the protection of the civilian population (see Jenks 2013: 108–126).

Targeting of individuals and the theatre in which this will take place are determined on a case by case basis. Determining elements include '…the imminence of the threat, the sovereignty of the other states involved, and the willingness and ability of those states to suppress the threat the target poses' (Koh 2010). Thus, official legal representatives like Harold Honju Koh (2010) have argued that targeted killing operations are '…conducted consistently with law of war principles'. Particular attention is paid to limiting attacks to military objectives (i.e., distinction) and ensuring that any incidental loss of civilian life, civilian injury, or damage to civilian objects are not excessive in relation to the military advantages conferred by a strike (i.e., proportionality). A leaked slide outlining the authorisation for targeted killing operations in Somali and Yemen claimed that doubts about the consistency of any operation with the laws of war, expressed by anyone in the kill chain, would result in a refusal of authorisation. Such a refusal would end the process (Currier 2015).

The American position has rejected the idea that the very act of targeting a leader is illegal by noting that leaders are belligerents under the laws of war and invoking past precedent, specifically the targeting of the Japanese architect of the Pearl Harbour attack during World War II. They have also argued that the laws of war are silent on the mode used to apply lethal force. As Koh (2010) stated, '...the rules that govern targeting do not turn on the type of weapon system used, and there is no prohibition under the laws of war on the use of technologically advanced weapons systems in armed conflict – such as pilotless aircraft or so-called smart bombs – so long as they are employed in conformity with applicable laws of war'.

In relation to objections that the US targeted killing programme is a form of extra-judicial killing, a direct appeal is made to position the practice as an action taking place within an environment defined as a war (e.g., Department of Justice 2013). Thus, the argument is that '...a state that is engaged in an armed conflict or in legitimate self-defense is not required to provide targets with legal process before the state may use lethal force' (Koh 2010). But while such a defence may justify the use of lethal force against a non-citizen, subsidiary legal justifications were provided by Holder for the question of the legal permissibility of targeting American citizens for elimination. The US government has claimed that US citizenship does not make an individual immune from targeting. Primarily, this claim rests on past precedent, including actions undertaken during World War II. However, the rationale concedes that all relevant constitutional protections offered to American citizens must be taken into account in such cases, particularly the due process protections to one's life offered under the Fifth Amendment. But, it is argued that Supreme Court decisions have shown that Due Process provisions are contextual. Thus, the private interest of the targeted individual for due process should be weighed in relation to the costs to the government of initiating all forms of due process. In the case of targeted killing, this means weighing private interest (for all aspects of due process) in relation to a claim that national security is threatened by an imminent violent attack for which there are no other viable measures of prevention.

Holder (2012) forcefully presented the Obama Administration's legal reasoning for the permissibility of targeting US citizens when he stated:

> Let me be clear: an operation using lethal force in a foreign country, targeted against a U.S. citizen who is a senior operational leader of al Qaeda or associated forces, and who is actively engaged in planning to kill Americans, would be lawful at least in the following circumstances: First, the U.S. government has determined, after a thorough and careful review, that the individual poses an imminent threat of violent attack against the United States; second, capture is not feasible; and third, the operation would be conducted in a manner consistent with applicable law of war principles.[7]

This rationale was further refined in *U.S. Policy Standards and Procedures for the Use of Force in Counterterrorism Operations Outside the United States and*

Areas of Active Hostilities which added additional criteria for initiating a strike:

> near certainty that the terrorist target is present; near certainty that non-combatants will not be injured or killed; an assessment that capture is not feasible at the time of the operation; an assessment that the relevant governmental authorities in the country where action is contemplated cannot or will not effectively address the threat to U.S. persons; and an assessment that no other reasonable alternatives exist to effectively address the threat to U.S. persons.
>
> (The White House 2013)

The standards and procedures document also affirmed that:

> whenever the United States uses force in foreign territories, international legal principles, including respect for sovereignty and the law of armed conflict, impose important constraints on the ability of the United States to act unilaterally – and on the way in which the United States can use force. The United States respects national sovereignty and international law.
>
> (The White House 2013)

Whether one finds the legal reasoning – or identity performatives – credible is not important.[8] Whether targeted killing is ultimately legally permissible – and who is legally empowered to commit it – is also somewhat besides the point.[9] What is of primary importance is how the cases for targeted killing have been made.[10] And the justifications have deployed a clear legal rationale, even if this has maximised state advantage by using various ambiguities within the law (e.g., proportionality) to gain room to manoeuver. Thus, there is not a normative suspension. Similarly, the US government stresses its adherence to rules and their positive application rather than primarily emphasising that rules need not apply in specific cases from first principles. If this is indeed a form of politics, it primarily represents a situation where the law and violence are indivisible from one another.

Conclusions

The perspective taken in this chapter demonstrates how relations of power shape the way in which assassination and targeted killing can be problematised and understood within legal regimes. In addition, the very tactics of counter-insurgency itself are actually contributing to legal problematisations of targeted killing and assassination. Emerging tactical options offered by RPA technologies may have opened new legal questions with respect to the status of pre-emptive killing as well as territorial considerations with regard to aggression as defined by the UN Charter but, just the same, it is argued that traditional legal principles are sufficient to assess these issues (e.g., Boyle

2015; Carvin 2015; Martin 2015). What thus becomes readily apparent is the inherent indeterminacy – or un-decidability – of the legal principles said to be regulating targeted killing as a field. The advantage of indeterminacy being left indeterminate is that what is permissible can always be subject to reformulation, reframing, and ultimately extension without necessarily resting on a definitive ground; this ambiguity can be used to stabilise targeted killing assemblages by maintaining spaces to operate within the law. The law therefore has a built-in flexibility in representing acts of assassination and targeted killing. And this flexibility lies at the very root of international jurisprudence: the evidentiary burden for the use of force. As O'Connell (2002: 21) notes 'despite over 100 years of international adjudication, and sixty years of Security Council fact-finding, we cannot point to any well-established set of rules governing evidence in international law in general or in the case of self-defence in particular'.

At the same time, this has not meant that legal considerations are ignored. In fact, some of the most vociferous proponents of the legality of targeted killing uphold the illegality of assassination. For those like David (2003: 115), the issue is that the substantial legal differences between assassination and targeted killing have been conflated in the minds of the global public. For him, 'assassination is a weapon of the weak. It benefits those with limited resources but fanatical devotion to a cause…it plays to Palestinian strengths' (David 2003: 116). He argues that it is actually important to maintain the international customary law prohibiting assassination and to differentiate targeted killing from it rather than letting the force of the ban erode. Israel and the United States have adopted this line of reasoning. More broadly, a legal analysis of assassination and targeted killing provides an alternative perspective on who we understand as the referent objects – that is the subject – to be protected under laws governing them. While international humanitarian law is often portrayed as a means to spare – as much as is reasonably possible – civilians from the horrors of war, the analysis above indicates that it is particular sets of combatants and those who command them who receive considerable protection. In particular, the ability of military commanders to escape legal liability through procedural – as opposed to substantive – means is endemic to considerations of anticipatory self-defence and proportionality. Similarly, the legal prohibition on assassination is one that has been established for those who are recognised as legitimate armed combatants not civilians. Being sensitive to who the law protects provides a different perspective on how modern warfare and counter-insurgency have been able to become institutionalised.

The analysis has also shown that international human rights law offers protections but that these too are flexible. In instances where the rules appear iron-clad – for example with respect to extrajudicial killing – there always exists the possibility to frame a plausible legal argument by changing the definition of the specific context in order to make a different legal framework apply. Again, this should not be viewed as an abnormality of the law or as evidence of its failings. Rather, this once again demonstrates how the law is both a progenitor and off-spring of political violence. This analysis thus forwards a critique of positions that might wish to understand the possibility of targeted killing through the exception as normative suspension. Moreover, while the interplay of sovereign authority, judicial process, and the law is important to understanding how targeted killing has been problematised, there would appear to be at least the residue – if not more – of a normative-legal basis for the claims being made. Thus, to label instances of targeted killing as a political act as conceptualised by Schmitt would discount the way in which legal justifications are presented by state authorities as already presiding within existing legal frameworks.

Agamben's (1998) conceptualisation of the exception is perhaps more interesting. The ambiguity of the sovereign ban does feature in the field of problematisation, perhaps most explicitly with regards to determinations of how much due process is required to target American citizens or when the Basic Law no longer applies in the Israeli context. It is present in the legal principles discussed above like proportionality or even war. Moreover, it may also be present in decisions made about what level of investigation is required into possible breaches of the law in cases of targeted killing. However, at least formally, the legal problematisation demonstrates that there is some legal form even if this is open to lawfare – i.e., being used strategically by state authorities to justify acts of extermination. Of particular note here is the firewall between acts of assassination and acts of targeted killing. The law certainly signifies, though the sign, as will be shown in the next chapter, may be located beyond the legal realm.

Thus, the exception as an analytic is not particularly useful beyond its focus on the interplay between sovereign authority and the law. In addition to the empirical inadequacy of the normative suspension assumption, exceptionalism evacuates politics, economics, and culture from the analytic frame. In fairness, collapsing the socio-political, economic, and cultural through the exception was Schmitt's political project. However, as Huysmans (2008) suggests, Agamben, in critiquing Schmitt, ends up in a similar space where social contestation and politics have been evacuated in lieu of a decisionism that operates within a vacuum. Both views of the exception then are unable to account for how targeted killing is mediated and obscure where politics takes place. The proceeding analysis has thus hopefully met Huysmans' (2008: 179) call to '…insert questions of and challenges to the role of law and generalised norm-setting in highly charged biopolitical governance of insecurities…[legal mediations of politics and life] open up a need to revisit the particular kind of work that law does and does not do in specific sites'. And this work is constituted and conditioned by diverse elements that fall outside of constitutionalism, the law, and even direct sovereign authority. As will be shown in the chapters that follow, narratives, regimes of truth, aesthetic subjects, and understandings of space that are all culturally conditioned contribute to the politics of targeted killing. As Derrida (1989/1990: 1045) suggested, 'if there were to be a lesson to be drawn, a unique lesson, among the always singular lessons of murder…it is that we must think, know, represent for ourselves, formalize, judge the possible complicity between all...discourses and the worst...'. In the analysis that follows in subsequent chapters, these lines of complicity in the targeted assemblage will be mapped out.

Notes

1 In theory, the legality of any particular act of targeted killing in war rests on how well it satisfies the principles of military necessity, discrimination (between combatants and civilians), and proportionality (i.e., that any harms to civilians are proportionate to the potential military advantages of a particular act).
2 While many popular discussions focus on drone strikes, the legal issues with respect to targeted killing are the same whatever the method of elimination. See for example, Carvin (2015).
3 These discussions will take place at the level of legal principles as opposed to a detailed analysis of the specific case law underpinning them. For a more detailed examination of the case law, see Melzer (2008).
4 This is taken up in more detail in the next chapter.
5 See Article 57 (2b) of the 1977 Geneva Conventions.
6 Again, it is worth noting that Israel has not signed the Additional Protocol to the Geneva Convention that presents key definitions around combatant and non-combatant status.
7 See also the Department of Justice White Paper (2013).
8 For critiques of this legal reasoning, see Boyle (2015).
9 The CIA is classified as a non-military organisation in international law. Thus, targeted killings enacted under its auspices, whether committed by agents or private contractors, could be argued to be at odds with rules regarding who constitutes a lawful combatant.
10 In 2015 after it was revealed that the UK government had targeted and killed a British national who was allegedly a member of ISIS, similar legal justifications to those used by the US were offered.

Bibliography

Agamben, G. (1998) *Homo Sacer: Sovereign Power and Bare Life* (D. Heller Roazen, Trans.), Stanford, Stanford University Press.

Ashkouri, M. (2001) 'Has United States Foreign Policy Towards Libya, Iraq & Serbia Violated Executive Order 12333: Prohibition on Assassination?', *New England International and Comparative Law Annual*, 7(1), 155–175.

Banks, W. C. (ed) (2013) *Counterinsurgency Law: New Directions in Asymmetric Warfare*, Oxford, Oxford University Press.

Bazan, E. B. (2002) *Assassination Ban and E.O. 12333: A Brief Summary.* Washington: Congressional Research Service, The Library of Congress.

Boyle, M. J. (2015) 'The Legal and Ethical Implications of Drone Warfare', *The International Journal of Human Rights*, 19(2), 105–126.

Braun, M. and Brunstetter, D. R. (2013) 'Rethinking the Criterion for Assessing CIA Targeted Killings: Drones, Proportionality, and Jus Ad Vim', *Journal of Military Ethics*, 12(4), 304–324.

Brooks, R. E. (2004) 'War Everywhere: Rights, National Security Law, and the Law of Armed Conflict in the Age of Terror', *University of Pennsylvania Law Review*, 153(2), 675–761.

Byman, D. (2006) 'Do Targeted Killings Work?', *Foreign Affairs*, 85(2), 95–111.

Carvin, S. (2015) 'Getting Drones Wrong', *The International Journal of Human Rights*, 19(2), 127–141.

Chamayou, G. (2012) *Manhunts: A Philosophical History* (S. Rendall, Trans.), Princeton, Princeton University Press.

Crona, S. J. and Richardson, N. A. (1996) 'Justice for War Criminals of Invisible Armies: A New Legal and Military Approach to Terrorism', *Oklahoma City University Law Review*, 21(2/3), 349–408.

Currier, C. (2015) 'The Kill Chain'. *The Intercept*. 15 October. Available from: https://theintercept.com/drone-papers/the-kill-chain. Accessed 15 October 2015.

Damrosch, L. F. (1989) 'Covert Operations', *The American Journal of International Law*, 83(4), 795–805.

David, S. R. (2003) 'Israel's Policy of Targeted Killing', *Ethics & International Affairs*, 17(1), 111–126.

Department of Justice (2013) 'Lawfulness of a Lethal Operation Directed Against a US Citizen Who is a Senior Operational Leader Al Qa'ida or an Associated Force', *Department of Justice*. Available from: http://www.cfr.org/terrorism-and-the-law/department-justice-memo-lawfulness-lethal-operation-directed-against-us-citizen-senior-operational-leader-al-qaida-associated-force/p29925. Accessed 15 October 2015.

Derrida, J. (1989/1990) 'Force De Loi: Le Fondement Mystique De L'Autorite/ Deconstruction and the Possibility of Justice', *Cardozo Law Review*, 11(919), 920–1045.

Dinstein, Y. (2005) *War, Aggression and Self-defence*, Cambridge, Cambridge University Press.

Downes, C. (2004) '"Targeted Killings" in an Age of Terror: The Legality of the Yemen Strike', *Journal of Conflict and Security Law*, 9(2), 277–294.

Dunlap, C. J. (2009) 'Lawfare: A Decisive Element of 21st Century Warfare?', *Joint Force Quarterly*, 54(3), 34–39.

Falk, O. (2014) 'Permissibility of Targeted Killing', *Studies in Conflict & Terrorism*, 37(4), 295–321.

Fletcher, G. P. (2006) 'The Indefinable Concept of Terrorism', *Journal of International Criminal Justice*, 4(5), 894–911.

Franck, T. M. (2004) 'Preemption, Prevention and Anticipatory Self-Defense: New Law regarding Recourse to Force', *Hastings International & Comparative Law Review*, 27(3), 425–436.

Grayson, K. (2012) 'Six Theses on Targeted Killing', *Politics*, 32(2), 120–128.

Gross, E. (2001) 'Thwarting Terrorist Acts by Attacking the Perpetrators or Their Commanders as an Act of Self-Defence: Human Rights Versus the State's Duty to Protect its Citizens', *Temple International and Comparative Law Journal*, 15(2), 195–246.

Gross, M. L. (2006) 'Assassination and Targeted Killing: Law Enforcement, Execution or Self-Defence', *Journal of Applied Philosophy*, 23(3), 323–335.

Guiora, A. (2004) 'Targeted Killing as Active Self-Defense', *Case Western Reserve Journal of International Law*, 36(1), 319–334.

Harder, T. J. (2002) 'Time to Repeal the Assassination Ban of Executive Order 12.333: A Small Step in Clarifying Current Law', *Military Law Review*, 172(1), 1–39.

Havens, M. C.Leiden, and K. M. Schmitt (1970) *The Politics of Assassination*, Englewood Cliffs, Prentice Hall.

Holder, E. (2012) 'Attorney General Eric Holder Speaks at Northwestern University School of Law', *United States Department of Justice*. Available from: http://www.

justice.gov/opa/speech/attorney-general-eric-holder-speaks-northwestern-university-school-law. Accessed 15 October 2015.

Huysmans, J. (2006a) 'International Politics of Insecurity: Normativity, Inwardness and the Exception', *Security Dialogue*, 37(1), 11–29.

Huysmans, J. (2006b) 'International Politics of Exception: Competing Visions of International Political Order between Law and Politics', *Alternatives: Global, Local, Political*, 31(2), 135–165.

Huysmans, J. (2008) 'The Jargon of Exception – On Schmitt, Agamben and the Absence of Political Society', *International Political Sociology*, 2(2), 165–183.

Jenks, C. (2013) 'Agency of Risk: The Competing Balance between Protecting Military Forces and the Civilian Population During Counterinsurgency Operations in Afghanistan' in W. C. Banks (ed.), *Counterinsurgency Law: New Directions in Asymmetric Warfare*, Oxford: Oxford University Press, 108–126.

Jones, C. A. (2015a) 'Frames of Law: Targeting Advice and Operational Law in the Israeli Military', *Environment and Planning D: Society and Space*, 33(4), 676–696.

Jones, C. A. (2015b) 'Lawfare and the Juridification of Late Modern War', *Progress in Human Geography*, DOI: 0309132515572270.

Koh, H. H. (2010) 'The Obama Administration and International Law'. Speech Delivered at the Annual Meeting of the American Society of International Law. (25 March). Available from: http://www.state.gov/s/l/releases/remarks/139119.htm. Accessed 22 February 2016.

Kramer, R. C. and Michalowski, R. J. (2005) 'War, Aggression and State Crime A Criminological Analysis of the Invasion and Occupation of Iraq', *British Journal of Criminology*, 45(4), 446–469.

Kretzmer, D. (2005) 'Targeted Killing of Suspected Terrorists: Extra-Judicial Executions or Legitimate Means of Defence?', *European Journal of International Law*, 16(2), 171–212.

Lotrionte, C. (2003) 'When to Target Leaders', *Washington Quarterly*, 26(3), 73–86.

Martin, C. (2015) 'A Means-Methods Paradox and the Legality of Drone Strikes in Armed Conflict', *The International Journal of Human Rights*, 19(2), 142–175.

Martinez, L. (2003) 'September 11th, Iraq and the Doctrine of Anticipatory Self-Defense', *University Missori-Kansas City Law Review*, 72(1), 123–192.

Melzer, N. (2008) *Targeted Killing in International Law*, Oxford, Oxford University Press.

Morrissey, J. (2011) 'Liberal Lawfare and Biopolitics: US Juridical Warfare in the War on Terror', *Geopolitics*, 16(2), 280–305.

Murphy, S. D. (2003) 'International Law, the United States, and the Non-military "War" against Terrorism', *European Journal of International Law*, 14(2), 347–364.

Neal, A. W. (2012) 'Normalization and Legislative Exceptionalism: Counterterrorist Lawmaking and the Changing Times of Security Emergencies', *International Political Sociology*, 6(3), 260–276.

O'Connell, M. E. (2002) 'Evidence of Terror', *Journal of Conflict and Security Law*, 7(1), 19–36.

Orr, A. C. (2011) 'Unmanned, Unprecedented, and Unresolved: The Status of American Drone Strikes in Pakistan Under International Law', *Cornell International Law Journal*, 44(3), 729–752.

Parks, W. H. (1989) 'Memorandum of Law: Executive Order 12333 and Assassination', *Army Law*, 4 (December), no page numbers.

Patterson, E. and Casale, T. (2005) 'Targeting Terror: The Ethical and Practical Implications of Targeted Killing', *International Journal of Intelligence and Counter-Intelligence*, 18(4), 638–652.

Paust, J. J. (2002) 'Use of Armed Force against Terrorists in Afghanistan, Iraq, and Beyond', *Cornell International Law Journal*, 35(3), 533–558.

Schmitt, C. (1985 [1922] *Political Theology: Four Chapters on the Concept of Sovereignty* (G. Schwab, Trans.), Cambridge, The MIT Press.

Schmitt, C. (1996) *The Concept of the Political* (G. Schwab, Trans.), Chicago, The University of Chicago Press.

Statman, D. (2004) 'Targeted Killing', *Theoretical Inquiries in Law*, 5(1), 179–198.

Strawser, B. J. (2010) 'Moral Predators: The Duty to Employ Uninhabited Aerial Vehicles', *Journal of Military Ethics*, 9(4), 342–368.

Thomas, W. (2000) 'Norms and Security: The Case of Assassination', *International Security*, 25(1), 105–133.

Welch, M. (2007) 'Sovereign Impunity in America's War on Terror: Examining Reconfigured Power and the Absence of Accountability', *Crime, Law and Social Change*, 47(3), 135–150.

The White House (2013) 'U.S. Policy Standards and Procedures for the Use of Force in Counterterrorism Operations Outside the United States and Areas of Active Hostilities'. White House. 23 May. Available from: https://www.whitehouse.gov/the-press-office/2013/05/23/fact-sheet-us-policy-standards-and-procedures-use-force-counterterrorism. Accessed 15 October 2015.

Zengel, P. (1991) 'Assassination and the Law of Armed Conflict', *Mercer Law Review*, 134(1), 615–644.

3 The politics of targeted killing
Introduction

In the previous chapter, I argued that the legal field contributes to processes of territorialising the targeted killing assemblage. It does so by stabilising its status as permissible in relation to assassination and by cultivating and maintaining the spaces where it can be committed. I concluded though that a narrow focus on the legal field, constituted as a battle between sovereignty and the law, overlooks key elements of the assemblage that are constitutive of its political effects. Primarily, the neglect of culture as a specialised expressive medium misses key dynamics contributing to the variable processes that are central to the assemblage and its future potentialities. The first are cultural values, narratives, and modes of interpretation that territorialise, either by solidifying the assemblage or producing flexible tolerances that allow for its perpetuation. The second are cultural values, narratives, and modes of interpretation that initiate sequences that could destabilise the assemblage by calling into question core components or relations. As argued in the opening chapter, problematisations are a key variable process. Thus, building upon the previous discussion of legal regimes, this chapter asks, how do culturally produced problematisations about what we may do to others make targeted killing politically possible in contemporary liberal regimes? At the same time, attention is paid to the fragility and ambiguity of these conditions of possibility and the ways in which their contingency has been culturally managed.

Typically, the question of what makes targeted killing politically possible has been answered by identifying core tensions within liberalism as a form of governance. These tensions involve the potentially competing demands of negative forms of freedom associated with classical liberal ideals, the pastoral imperatives envisioned as necessary for the maintenance of liberal forms of governance, and the security of liberal subjects. As presented in the opening chapter, it has been argued that the extension of forms of knowledge from the life sciences into the governance of populations by liberal regimes has enabled the adoption of tactics, techniques and outcomes in the name of security that can be detrimental to the rights and freedoms of individuals and groups while seeking to govern in their interests. Biopolitics is thus positioned as an aspect of contemporary liberalism that both enables and eliminates forms of life. To do so, it often presents what might be perceived as excesses of force as exceptional measures undertaken to respond to existential threats.

I agree in principle with these assessments of what becomes possible within liberalism when existential threats are perceived to be producing insecurity for prescribed ways of life; however, my argument is that although biopolitical accounts provide important – and competing – insights into the processes that turn individualised killing into a mechanism of security provision, it is also important to account for how it has become possible for assassinations and targeted killings to register as politically significant events beyond their positioning as *exceptional* measures within security discourses. To address this limitation, I begin the analysis in this chapter by presenting a biopolitical account of assassination and targeted killing by liberal regimes. I link this account to 'cynegetic' power identified by Gregoire Chamayou (2012). I then argue that these biopolitical explanations overlook cultural elements that position their mechanisms within political narratives as a particular kind of high-profile *event*. Deploying Alan Feldman's (1991) argument that political violence is an 'emplotted action' with William Connolly's (2005) notion of resonance, I provide a broader genealogical account of how forms of assassination have been placed within Western cultural narratives of political violence. My focal point of examination is the biblical heroine Judith whose story has resonated across various historical contexts as a preferred narrative structure for understanding and (de)legitimating acts of assassination. From my reading of the Book of Judith, the importance of ambivalence to the dynamics that allow for acts of assassination and targeted killing to be politically legitimated is highlighted. I then demonstrate how this ambivalence has become manifest in two high profile cases of targeted killing: Mahmoud al-Mabhouh and Osama Bin Laden. I conclude by suggesting that ambivalence provides an opportunity to challenge the security (bio)politics of targeted killing. Thus, the turn to the drone may function, in part, as a way of distancing targeted killing from assassination.

Assassination and targeted killing as biopolitical narratives of violence

In the uncovering of deviations from the norm and their identification with dangerous populations and individuals, biopolitics engenders a wide spectrum of interventions and acts of violence. The focus is often on the most overt of these practices. As an analytic framework, biopolitics draws attention to how forms of subjectivity – both for individuals and at the level of population – are inscribed with characteristics that determine the institutions, mechanisms, processes, and types of knowledge that will be applied to governing their lives (Dean 1999: 15). It also remains sensitive to how problematisations themselves designate preferred forms of intervention. But, as Michael Dillon (2007: 45) has noted with respect to understanding the characteristics and behaviours of human beings that are implicated in the construction of problematisations:

> in governmental terms, the contingent features that life and populations display…are a function of truth-telling practices of the life sciences,

uncertainty, and risk. These perform a whole variety of governmental as well as scientific functions, not least in telling different stories about different categories of living things and their governability, as well as what falls into the category of living things as such.

As mentioned in the introduction to this monograph, in order to govern, one needs a population that is governable. To be able to determine when and when not to intervene, liberalism requires biopolitics. And the major questions that define how to govern effectively, that is to (re)produce populations who can be governed within the confluence of liberalism and biopolitics are: *who must be free to live and/or to die, who must be monitored, who must be controlled, who must be corrected, who must be punished, and who must be killed* (Dillon and Reid 2009: 87)?

In attempting to understand what makes possible the most overt and coercive interventions that these questions inculcate, analyses concerned with the biopolitical dynamics of contemporary liberal governmentality locate the meanings that can be attributed to acts of violence enabled by these problematisations within that governmentality itself. Violence that conforms to the regulatory norm thereby helping to maintain 'species life' is said to reflect dispersed administrative decisions and forms of power-knowledge (e.g., statistical sciences, biomedicine, insurance) that may inform decision making to manage the threats said to emerge through the contingency of life (e.g., Dillon and Reid 2009: 93). Violence as conceived within biopolitical accounts has therefore been presented – and critiqued – as de-personalised, bureaucratised, and ultimately de-politicised in the cold calculating manner lamented by early twentieth-century social theorists like Max Weber (2005), Walter Benjamin (1986), and C.W. Mills (2000). With the epistemology of pre-emption structuring problematisations of (in)security across political assemblages, these practices, what they are said to be pre-empting, and their effects become difficult to counter-act within the grammar of security discourses obsessed with managing risk (see Kessler and Werner 2008; Massumi 2007; De Goede 2008; Muller 2008).

What emerges from these processes has been identified as a paradox of liberal governmentality and a shared element of some contemporary problematisations of violence: in the provision of social protection, we all have the potential to become *homo sacer*, the body that can be killed without registering as either murder or sacrifice (Agamben 1998). It is argued that we are all capable of being 'bare life' because we are all infinitely expendable under the right matrix of risk in the right circumstances (Jabri 2006). The rationale can be based on an inherent condition – a 'congenital defect' – or a mutation. It may also be the result of adapting through cybernetic interaction – with an environment or other stimuli – into an undesirable or ungovernable form. From a perspective sensitive to power-relations engendered by biopolitics, targeted killing and assassination from this perspective can be seen as complex and potentially contradictory responses. On the one hand, they are the individualisation of danger to the extreme in that an existential threat to a way of life is located – at some moment in time – within the capabilities and/ or intentions of a single person. On the other hand, the act is not personal; the administrative decision to despatch is not so much a result of 'who you are' as 'what you are' or 'what you have been determined to be': an existential threat to the survival of the species.[1]

Biopolitical logics and the violence they may engender are thus connected to other power analytics that have been identified by Foucault – sovereign, juridical, and disciplinary – as well as relations of control outlined by Gilles Deleuze (1992). They produce assemblages of power in which these forms mobilise, interact, and even contest practices of political economy, justice, care, circulation, territory, and security that shape the deployment of violence. For example, an act of violence can be predicated on biopolitical calculations of risk assessment. It can also be legitimated under legal doctrines that give sanction to both the tactic and sovereign decision to implement it. The same act of violence may be publicly enacted with spectacular force. The intention may be for its corrective message to be internalised by other potentially at risk subjects such that problematic behaviours are prevented from manifesting through self-imposed forms of discipline. Moreover, the form of violence itself may have a symbolic force and significance that goes beyond what might be inferred by its constitutive power-relations. As such, it is important to specify the relations of power underpinning violence rather than subsuming every-thing under biopolitical rationalities said to be fostered under liberalism. Such specificity is important to understanding the cultural politics of assassination and targeted killing.

Manhunts: an opening into the cultural politics of assassination and targeted killing

One possible means of providing additional specificity to the relations of power underpinning assassination and targeted killing is to more precisely locate the governing logics underpinning them. I wish to take it as read that contemporary forms of violence can be connected to biopolitical rationalities under liberalism. That being said, these connections, their strength, and other relevant factors – including additional power dynamics – will be dependent on the type of violence, the context in which it is deployed, the perpetrator(s), and the recipient(s) – including how these elements are defined. While the logics of biopolitics may inform decisions about who to target – and for what reasons – the form and purpose of the acts of violence may reflect longer-standing (cultural) practices that signify relations of power. In *Manhunts: A Philosophical History*, Grégoire Chamayou (2012) demonstrates that cyne-getic relations of power (i.e., those social relations that stem from hunting other human beings) are an important factor in understanding contemporary governing practices from policing strategies to locating undocumented migrants. His central argument is that '...the technologies of predation [are] indispensable for the establishment and reproduction of domination' (Chamayou 2012: 1). To support this contention, Chamayou provides a genealogical analyses of how manhunting – and the rationalities underpinning it – have developed over time.

With respect to biopolitics, Chamayou (2012: see chapter 3) shows how Christian pastoral power and cynegetic power began to coalesce during the Middle Ages as authorities in Europe became increasingly concerned with socio-political threats believed to originate from outside of the established spiritual order. Table 3.1 provides a brief overview of key distinctions between pastoral and cynegetic forms of power. Pastoral power was trascendental and focussed on the people as a flock – traversing within the boundaries of a 'pasture' – who required the spiritual guidance of a shepherd in order to reach salvation. Cynegetic power was terrestrial and bounded by a secular temporality of the here and now. The task was not salvation but the pursuit and accumulation of people who were defined as prey. Where pastoral power concerned itself with the spiritual health and well-being of its flock, including the formation of individual subjectivities, cynegetic power intervened for the purposes of accumulation, that is to tax and consume people as property.

When these two forms were brought together initially in response to perceived threats to the Christian order in Europe, it resulted in pastoral hunts. The logic was not the accumulation of prey for slavery or the capture of wayward subjects for punishment. Rather, Chamayou (2012: 22) argues that pastoral hunts operated under the logic of exclusion: they '…tend to eliminate the subjects

Table 3.1 Pastoral power and cynegetic power

	Pastoral power	*Cynegetic*
Source	Transcendental (spiritual)	Terrestrial (temporal)
Object of attention	Flock	Prey
Form of mobility	Guide	Pursuit
Geographical imagination	Pasture	Space (capture) and territory (accumulation)
Power form	Beneficent	Predatory
Focal point on subject	Individualisation	Accumulation
Modality of intervention	Concern over life and health of subjects	Taxation and consumption (to the death)
Governing tension	Whole v. part (preservation v. sacrifice)	None: 'all can die'
Typology	Christian, biopolitical, liberal	Acquisition hunts (slavery), capture hunts (tyrannical sovereignty), exclusion hunts (pastoral power/ biopower/liberal)

concerned'. The point is not to acquire additional prey but to dispatch something (or someone) that has become perceived as dangerous. Beyond the provision of security for an order through the elimination of threats, there is also an ontological security that is provided through practices of manhunting. In his discussion of the Spartan *crypetia*, Chamayou (2012: 10) argues that the hunting and killing of helots was '...a means of *ontological policing*: a violence whose aim is to maintain the dominated in correspondence with their concept, that is, with the concept the dominant have imposed on them'.

This interplay of violent practices such as pastoral hunts and imposed subject positions became further entrenched with colonialism. Thus, Chamayou equates colonialism with the rise of a new form of warfare: cynegetic war. Unlike traditional forms of warfare, cynegetic war involves processes of tracking rather than a direct confrontation with opposing forces. Its power-relations are marked by a stark imbalance in weapons capabilities with hunters greatly advantaged over their prey. While involving two main parties, the structure of violence is not a dual between adversaries but a largely unidirectional application of force. In this struggle, the enemy is not recognised as an equal agent but as prey to be dictated to and imposed upon. And rather than using standard military tactics, cynegetic warfare adopts policing and hunting techniques to capture and eliminate opposition (Chamayou 2012: 73). This is the form of warfare that underpins the practices of surveillance, capture, and elimination that constitute the wars of what Derek Gregory (2004) has called 'the colonial present': the war on drugs; the war on terror; the wars against illegal migration; the wars of pacification; and wars of counter-insurgency. The hunt provides a common matrix of rationalities and practices, including targeted killing, for producing relations of domination.

Therefore, what Chamayou's argument suggests is that the resurgence in targeted killing events organised by the state as 'manhunts' is not something exogenous to liberal forms of rule – bearing in mind that it is not exclusive to liberal regimes either – or a failure in their capacity to curb the excesses of security biopolitics. Rather targeted killing can be a central element to liberal forms of rule when confronted with what are understood as particular types of threat arising from people who are understood in particular ways. This is one of the key contributions of Chamayou's analytic: it draws attention to who is subject to the manhunt and its associated forms in a given socio-political context. More importantly, as Aarons (2013) has argued in relation to his previous work on medical experimentation, Chamayou neither positions this logic as an external political imposition onto practices of governing nor does he reduce it to ethical considerations internal to the practices of governing. Rather, he shows how manhunting is central to the composition of governing itself. It is an essential task of maintaining order: the functional separation of predator from prey.

Within the context of global counter-insurgency and chaoplexic war-fighting where networks feature prominently, Crawford (2009: 1) has argued that manhunting now marks '...the deliberate concentration of national power to find, influence, capture, or where necessary kill an individual to disrupt a human network'. High value targets for manhunting are said to be determined through the collection of precise intelligence – both electronic and human-based – regarding capabilities and intentions as well as analyses that identify key nodes in insurgent networks.[2] But at the same time, it would be incorrect to assume that what shapes these targeting determinations is relegated absolutely to the directives of liberal governmentality, and/or to the dictates of 'life as code', and/or to the politics of the exception, and/or cynegetic impulses, and/or the limits of reflexive modernity (see Dillon and Reid 2009; Agamben 1998; Chamayou 2012; Beck, 1992; Williams, 2008). For example, beyond their political economy, Chamayou's (2012) analyses reveal the ritualistic aspects of exclusion hunts and the cultural belief systems underpinning them. Similarly, Carvin and Williams (2014) have attributed the resurgence in targeted killing to the institutional culture of the US military. Even Dillon and Reid (2009: 6, 9) note that cultural representations must play some role in contemporary biopolitics, though they neglect to specify what role that might be.

Culture, politics, and violence

Although the deferral of culture may just be an analytic convenience to deal with the complexities of a 'palimpsestuous phenomenon' where several elements contributing to governance can be found in traces on the same page, it risks compartmentalising politics and culture rather than exploring their mutual constitution (Dillon and Reid 2009: 84–85). For Foucault cultural practices like confession and the Christian pastoral ethos were not only central to the development of specific forms of individual discipline and biopolitical management but also important to our experiences of desire, pleasure, and care (Foucault 1990; 2007). Moreover, Foucault (1990: 42–43) revealed how individuals who may have previously been subject to scrutiny and probation under Judeo-Christian moral codes (e.g., the sodomite) were transformed from 'juridical subjects' into 'a type of life, a life form...' through the development of the human sciences. Homosexuals – and other populations of 'interest' – then became repositioned as social dangers requiring a whole host of interventions to protect the *health* of the *body* politic as opposed to remaining subjects who merely required juridical sanction. Cultural prohibitions were thus intertwined with the development of biopolitical management and contributed to the formation of emerging predator–prey relationships.

This cultural process of configuring biopolitical control is also a theme in post-colonial explorations of power. Both Edward Said (1978) and Homi Bhabha (2005: 122) identify the 'conflictual economy' between identity-stasis (i.e., that certain populations were inherently inferior and unable to 'advance') and difference-change (i.e., a desire to rehabilitate and Europeanise populations) as a defining tension of colonial projects. Cultural chauvinisms on the part of colonisers and forms of cultural mimicry undertaken by the colonised (e.g., religious conversion) were thus able to generate oscillations in perceptions of colonial subjects from 'almost the same, but not quite' to the menacing 'almost total, but not quite' (Bhabha 2005: 131). Concurrently, the perceived space of difference was continuously plotted along a continuum that ranged from disgust to desire.

Culture practices themselves are neither static nor necessarily internally coherent. Bhabha (2005: 95, 129) thus argues that unstable notions of cultural difference can be transposed into biopolitical forms of management that attempt to concretise notions of race as distinct categories with established characteristics that are 'in *excess* of what can be empirically proved or logically construed'. These categorisations and their associated hierarchies help to constitute the forms of violence undertaken in the name of humanity that follow. More importantly, the initial perceptions of cultural difference, as plotted within the conflictual economy of identity and its affective continuum, shape the scope, intensity, and types of violence that may be initiated. It is these notions of cultural difference that can provide additional latitude for killing.

Drawing from the work of Chamayou, Foucault, and Bhabha would then suggest that while biopolitics may provide an administrative framework for contemporary liberal governmentality, the narrative structure of its understanding of violence is more dispersed (see De Goede 2008; Huysmans 2008: 177–180; Pan 2009). The point is to move beyond a limitation that biopolitical analyses can inadvertently share with orthodox forms of security studies: a narrow focus on technical rationalisations for action at the expense of exploring how these rationalisations – and biopolitics – are also culturally produced and circulated. The interest here is with how (liberal) biopolitics as a political culture may draw upon more than technical rationality in its deployment of violence and in its own self-understanding. At the same time, it will be shown that having a broader narrative to orient violent events does not necessarily remove anxiety over the legitimacy of forms of political violence, even if they are congruent with the selective logics of biopolitics. The importance of culture is apparent when examining assassination and targeted killing.

First, as Campbell (1998), Weldes et al. (1999), and Hansen (2006) have demonstrated, broader political cultures are important to the construction of threats. With regards to assassination and targeted killing events, Israel and the United States can be identified as two of the most prominent contemporary cases amongst regimes considered to be liberal. It is perhaps unsurprising that both states have deep histories of orchestrating manhunting and assassination events. In the armed struggle for the establishment of an independent state of Israel, Ben-Yehuda (1990: 363) documents approximately 87 assassination attempts were orchestrated by Jewish separatists within Palestine from 1918 to 1948 with the peak period occurring from 1937 to 1948 during increased activity against the British occupation. Members of the insurgency often went on to careers within the government and security forces of Israel. Away from the independence struggle, others who would become members of the security forces for the fledging state of Israel were former members of the *Hanokmin* (Avengers), a paramilitary group formed from the Jewish Brigade in the

British military. In the immediate aftermath of World War II, the *Hanokmin* organised itself into a group for tracking down prominent Nazis, particularly members of the SS who were involved in the management of concentration camps. With a Europe-wide network of informants and trackers, including sympathetic collaborators in the occupying forces, the *Hanokmin* located and captured hundreds of targets. While the initial practice was to deliver suspects to Allied authorities for prosecution, inadvertent releases of former Nazis in the chaos of post-war occupation led to a change in tactics: immediate execution upon capture. Eisenberg et al. (1978: 24) report over 1000 such killings took place in the first year after the end of the war.

As Eisenberg et al. (1978) document, from the 1950s onwards, Israel received international attention for its capture of Adolf Eichmann in Argentina and alleged involvement in multiple assassinations of German scientists contributing to Egypt's burgeoning rocket development programme. In the aftermath of the killing of Israeli athletes during the Munich Olympics (1972), the nine men who the Israeli authorities believed were associated with Palestinian terrorist acts in various European countries as well as members of Black September and the PLO in Beirut were executed. This particular set of missions was facilitated by the establishment of the *Mivtzan Elohim* – or Wrath of God – a unit organised to commit assassinations abroad. In the 1970s, the killing of a waiter in Lillehammer, Norway – through misidentification – caused the pace of Israeli operations abroad to subside; however, targeted killings committed by agents of the state of Israel did not stop and continued on through to the present day. Jones (2015: 681) argues that the beginning of the second intifada in 2000 marked a new era for individualised killings as their high profile within the conflict '...required an extensive legal apparatus, including the deployment of military lawyers in the planning and execution stages of these operations'.

Likewise, Crawford (2009) has argued that manhunting and targeted killing undertaken on behalf of the government have been established in the United States since colonial times. These practices were often deployed in those spaces that formally marked American territory but that were *de facto* subject to weak control such as frontier states. Manhunting was also directed against populations perceived to be a threat – whether imminent or latent – such as indigenous peoples and slaves. Wanted dead or alive posters, bounties, contending branches of law enforcement, and formations of quasi-legal 'posses' contributed to formal and informal practices of manhunting. As Richard Slotkin (1973; 1992) has argued, American political culture was shaped by these experiences and their later representation through literature and film. Similarly, practices associated with manhunting, such as lynching, continued well into the latter half of the twentieth century, and harkened back to the spectacular forms of violence inflicted upon slaves (Garland 2005). In his extensive study, Crawford (2009: 2) is able to demonstrate at an official level how manhunting and the targeting of political leaders by the American government has accelerated over the past three decades. Examples here would include the capture of Manual Noriega, the targeting of Momar Gaddafhi in 1986 and again in 2011, numerous attempts to capture and kill Sadam Hussein, as well as the extermination of Osama Bin Laden. Thus, manhunting and targeted killing have influenced the contours of the internal and external geopolitics of the United States.

Second, Dillon and Reid (2009: 38) have argued that the legitimacy of biopolitics is tied to notions of effectiveness. If this is true, 'effectiveness' must in turn rely on other forms of cultural representation in order to be comprehensible. Targeted killings then pose a problem for biopolitics' technical rationalisms and any account that wishes to emphasise their singular importance. First, the use of targeted killing can potentially be seen as a response to the failure of subjects to adequately self-regulate and/or for the prevailing governmentality to effectively anticipate and mitigate dangerous capabilities through less invasive controls. Second, publicly available academic research that has been undertaken – research that uses the same grammar as biopolitics – often demonstrates that targeted killing has had minimal impact on the frequency and/or scope of insurgent attacks (e.g., Hafez and Hatfield, 2006; Kober, 2007; Honig, 2007; Morehouse 2014).[3] Moreover, findings from a series of longitudinal statistical regressions has led Jenna Jordan (2009: 735–736) to conclude that not only is the overall decline rate (17 per cent) of organisations that have experienced the targeted killing of leadership less than those who had not had leaders removed through this practice, but also that decapitation strikes (i.e., targeting the leadership of these organisations for killing) are '… actually counter-productive against large, old, and religious groups'. In particular, Jordan (2014b) would later find that levels of bureaucratisation and communal support mitigated any potential disruption from the elimination of insurgency leaders.

Third, the probable ineffectiveness of targeted killing is actually foreshadowed in contemporary strategic discourses and the representations of terrorism that focus on the image of the network (Honig 2007: 571; Coward 2009; Croft and Moore 2010: 825–827). What is said to make groups like Al Qaeda, Izzedine al-Qassam (a branch of Hamas' military wing), and Hezbollah so dangerous is their largely decentralised operational structure that allows for individuals or cells to conduct operations as they see fit, rather than having to rely on orders or capabilities generated from a broader organisational structure (Bunker and Begert 2002: 329). In other words, killing individuals – particularly those identified as leaders – within these kinds of structures is not likely to have the anticipated effect or fulfil the desire of creating organisational disarray.

That assassination and targeted killing may be entirely ineffective at meeting stated security aims does not undermine arguments that stress the importance of biopolitics to liberal forms of governance. Similarly, ineffective policies do not necessarily signal that a biopolitical rationalisation is absent. But, the mounting evidence of targeted killing's limited effectiveness indicates that its policy resiliency must be drawing upon other logics and forms of rationalisation. Despite an acknowledgement that retribution has no justifiable basis in legal argumentation around targeted killing – in part because of the legal regimes governing armed conflict and those establishing human rights protections against extrajudicial killing – some arguments in favour draw upon a politics of retribution based on justifications that the practice provides targets with their just dessert (see interviews in Ben-Yehuda 1997; David 2003a; 2003b; Patterson and Casale 2005). Normative evaluation of this type is reliant on broader cultural representations that indicate that retribution (or revenge) is an appropriate deliberative framework as well as culturally embedded understandings of the individual targeted in order to reach such judgements. How then should we situate this cultural modality to best capture its mutual constitution with a mixture of biopolitical rationalities, processes, and mechanisms that make forms of political violence that register as assassination and targeted killing events possible?

Political violence, assassination, resonance

A sensitivity to the biopolitical dimension of liberalism reveals that political violence is intimately related to regimes of truth and power/knowledge that shape who is allowed to speak, the positions that can be acceptably articulated, the institutions that are able to serve as conduits of speech, and the institutions that store and distribute what is said (Foucault 1990: 11). But the 'truth' of political violence is more than the technical rationalities that may enable and justify biopolitical management. It is also present in the understandings, representational practices, and desires – i.e., forms of symbolic violence – that create political possibility. Often understandings of political violence and their associated meanings take the form of narratives, stories, and histories in order to make them comprehensible. To enable comprehension, Allen Feldman (1991: 14) argues that 'political violence is a genre of "emplotted action"... [an action of narration that organises] events into a configurational system, a mode of historical explanation, and a normative intervention.' Political violence as an emplotted action presents a temporal network that internally sequences actions/events through modes of self-referral and description aimed at a target audience (Ricoeur 1973: 92–93). Acts of political violence are thus events that are linked to other events, 'violent' or otherwise. But Feldman (1991: 14) reminds us that *the event is not what happens. The event is that which can be narrated.* The event is action organised by culturally situated meanings' as well as what can be articulated, and what can be understood within the dominant iterative of the day. Narratives [of political violence] are therefore a means of managing uncertainty over the meaning: they express a desire to order, capture, and fix a preferred reading of an event despite their own internal instability (Connolly 2011). Similarly, narratives are as much enacted as they are written (Feldman 1991: 14). In configuring mediations of experience, political violence as an emplotted action makes mimetic performances by others possible; for an assassination to be understood as an assassination, it must mimic the configurations, historical positionings, and normative interventions encapsulated in assassination narratives. As the more recent practice, targeted killing is positioned in key discourses as something different to assassination. Yet, as will be demonstrated in the analysis that follows, targeted killing can struggle to differentiate itself from assassination in terms of how it is emplotted in narratives. Thus assassination and targeted killing events are emplotted and articulated within culturally situated meanings that go beyond the technical rationalisations that may foster them.[4]

Events, political or otherwise, are not self-contained; understandings are shaped by latent sensibilities, representations, and scripts that can be drawn upon, combined, and distilled to modulate the affective intensity of our political sentiments. In the contemporary (geo)political landscape, William Connolly argues that these:

> ...diverse elements *infiltrate* into the others, metabolizing into a moving complex...heretofore unconnected or loosely associated elements *fold, bend, blend, emulsify, and dissolve into each other*, forging a qualitative assemblage resistant to classical modes of explanation.
>
> (Connolly 2005: 870; italics in original)

These dynamics of resonance therefore amplify and extend the constitutive effects of disparate elements within politics including emplotted actions. While Connolly's (2005: 879) specific concern is to challenge the 'evangelical-capitalist resonance machine' in American domestic politics, his identification of the power of the 'will to revenge' that 'reverberates back and forth between leaders and followers' within this assemblage is pertinent to understanding the politics of assassination and targeted killing. He links the 'will to revenge' to selective readings of the Bible undertaken by the religious right that privilege the punitive justice of the books of the Old Testament and/or Revelations. With these readings providing a familiar script, security practices like targeted killing serve as the echo chambers of mimesis through which the 'will to revenge' and biopolitical imperatives reverberate and recombine.

But there is also a moral ambivalence that resonates politically in assassination events. Beyond 'the will to revenge', assassination as a distinct type of political violence in liberal regimes is concurrently shaped by other Judeo-Christian representations of the practice. Thus, assassination is also emplotted as an ambivalent normative intervention. As such, targeted killing and assassination draw attention to the necessity to begin to think about the ways that broader representational practices contribute to, or shape, biopolitical rationalities and the policy postures that their resonance makes possible.

Biblical representations of assassination[5]

In terms of resonance, the most powerful story of assassination in the Bible is found in the Book of Judith, canonical in the Orthodox Church, deuteron-canonical in Catholicism, and Apocryphal in Judaism and Protestantism. While expressly recognised by leading theologians as a work of historical fiction, since its time of writing around 100 BC, it has provided a primary means of problematising assassination in Judeo-Christian societies (Stocker 1998). The text itself outlines the tale of the widow Judith of Bethulia and her methodical slaying of Holofernes, a general commanded by King Nabuchodonosor of Assyria. Holofernes has been ordered to violently punish the people of Israel for their unwillingness to worship Nabuchodonosor as their god. In the midst of an Assyrian siege of Bethulia – strategically important as the last line of urban defence before the holy city of Jerusalem – town elders prepare to surrender within five days unless given a sign that God will protect them. Judith, a widow recognised as one of Bethulia's most beautiful, chaste, and pious residents, admonishes the elders for having the audacity to hold God to a worldly schedule. Instead, she offers to address the problem on her own. After ritualistically cleaning and dressing in her finest robes, Judith and her maid travel to the Assyrian camp, offering to reveal secrets about Bethulia's wavering resolve if she can speak directly to Holofernes.

While waiting for an audience, she establishes a pattern of preparing her own food and fastidiously bathing to a regular schedule. Upon seeing Judith, Holofernes is smitten. After keeping his company and playing coy for several days, Judith attends a feast hosted by Holofernes who has now become hopelessly infatuated with her. Over the course of the meal, she encourages him to drink himself into a stupor. Once he is unconscious back in his tent, Judith strikes, decapitating him with his own sword. The details of the assassination are described as thus:

> Then she came to the pillar of the bed, which was at Holofernes' head, and took down his fauchion from thence,
> And approached to his bed, and took hold of the hair of his head, and said, Strengthen me, O Lord God of Israel, this day.
> And she smote twice upon his neck with all her might, and she took away his head from him,
> And tumbled his body down from the bed, and pulled down the canopy from the pillars; and anon after she went forth, and gave Holofernes his head to her maid;
> And she put it in her bag of meat: so they twain went together according to their custom unto prayer: and when they passed the camp, they compassed the valley, and went up the mountain of Bethulia, and came to the gates thereof.
>
> (The Book of Judith: 13:6–13:10)

By working within her established routine of regular bathing, she is able to leave the Assyrian camp with her maid unchallenged. Upon returning to Bethulia, she displays Holofernes' severed head, claiming that the killing and its circumstances – including the protection of her chastity against his licentious intentions – were directly facilitated by God to protect his chosen people. Her speech is described as thus:

> Then she said to them with a loud voice, Praise, praise God, praise God, [I say,] for he hath not taken away his mercy from the house of Israel, but hath destroyed our enemies by mine hands this night.
>
> So she took the head out of the bag, and shewed it, and said unto them, behold the head of Holofernes, the chief captain of the army of Assur, and behold the canopy, wherein he did lie in his drunkenness; and the Lord hath smitten him by the hand of a woman.
>
> As the Lord liveth, who hath kept me in my way that I went, my countenance hath deceived him to his destruction, and yet hath he not committed sin with me, to defile and shame me.
>
> Then all the people were wonderfully astonished, and bowed themselves and worshipped God, and said with one accord, Blessed be thou, O our God, which hast this day brought to nought the enemies of thy people.
>
> (The Book of Judith: 13:14–13:17)

The speech provides a justification for the slaying through Judith's deployment of security and retributive rationales. Although committed through deception, Judith claims that Holofernes' death by decapitation will not only save Israel from Assyrian domination but that it was also an act of divine justice. Judith's assassination of Holofernes is thus able to catalyse resistance to Assyrian rule. Upon discovering his mutilated body, Boagas, an Assyrian captain announces:

> These slaves have dealt treacherously; one woman of the Hebrews hath brought shame upon the house of king Nabuchodonosor: for, behold, Holofernes lieth upon the ground without a head.
>
> (The Book of Judith: 14: 18)

With news that their leader is dead, the Assyrian army becomes 'wonderfully troubled' and cries of anguish reverberate throughout the camp (The Book of Judith: 14:19). The will of the Assyrians to fight crumbles and they abandon the siege. In the ensuing chaos of the sudden retreat, the people of Israel counter-attack, slaughtering the Assyrian army and ending Nabuchodonosor's brutal occupation.

In one sense, the story of Judith is not entirely unique as there are other biblical tales where women play a decisive role in the killing of a dangerous adversary. In the Book of Judges, Jael, the wife of Heber, slays Sisera, former overlord of Israel and defeated general on the run, after offering him sanctuary in her home. Innuendo can be read into the levels of deference and submissiveness exhibited through her offers of hospitality (Wallhead 2001; Reis 2005; Mayfield 2009). Her apparent passivity is a ruse and Jael ultimately kills

Sisera while he is sleeping by driving a tent peg through his skull with a hammer. And for this act, later chapters reveal that Jael is to be revered.

Thus there is a gender politics underpinning these acts of political violence and the way that they were understood. Ford (1985: 11) argues that to be killed at the hands of a woman was considered a particularly undignified way to die by referencing the lesser known story of the Israelite tyrant Abimelech. In laying siege to the city of Thebez, Abimelech is mortally wounded when a woman drops a millstone upon his head from a high tower along the city wall, breaking his skull. The events that follow are recorded as thus:

> Then he called hastily unto the young man his armourbearer, and said unto him, Draw thy sword, and slay me, that men say not of me, A woman slew him. And his young man thrust him through, and he died.
>
> (The Book of Judges: 9:54)

The (gendered) representations in the Book of Judith and other biblical tales of assassination are important and were filtered and remixed through related social codes. Contemporary theologians and historical anthropologists have argued that at the time of its first recounting, the story of Judith would have been primarily recognised as a parable about shame and honour (Esler 2002; deSilva 2006). In this case, one could argue that God is attempting to preserve the honour of a chosen people, while Judith is attempting to preserve the honour of her God (deSilva 2006). Holofernes as an outsider ethnically, culturally, and spiritually is not to be accorded the respect of an honourable death on the battlefield at the hands of a man or men.

The Book of Judith from its first articulation was thus constitutive of a cultural system of configuration, forms of explanation, and a normative compass by which assassinations could be positioned as events. Given cultural codes, the technical form of death – beheading – was appropriate in that it befitted someone of the stature of Holofernes to be killed in this way and symbolically allowed 'access to a mystical form of empowerment' for Judith (Janes 2005: 11).[6] Yet, historically there has been a lingering unease about how this killing was undertaken. Although the rationale is argued to fit within prevailing ethical codes of the time, and its facilitation by God makes it just by definition within these codes, there is also an acknowledgement by the perpetrator to the audience that the act has involved deception, treachery, seduction, and devious premeditation. Furthermore, there is a sexual sub-text at play, with passing references to flirtation, attraction, and perhaps – as in later re-tellings – even illicit congress (see Ziolkowski 2009: 316–321).

But the assassination is not just facilitated by God at arm's length. Theologians over the centuries have shared an interpretation that divine intervention is central to this story, in particular that Judith herself was – to put it crudely – created for purpose by the hands of God to serve as an instrument of those very same hands, including her ability to push beyond traditional gender boundaries (Day 2001: 71–72; Sawyer 2001a: 30; Sawyer 2001b: 15). This is what distinguishes her from other 'deceitful' women in the bible like Eve or Delilah. Judith is revered, not just by the residents of Bethulia until her death, but also very often by those who read or have her story recounted to them over the ensuing centuries. For example, she is a paragon of Christian virtues – Chastity, Temperance, Justice, Fortitude, Wisdom, and Humility – in the *Psychomachia* written by Prudentius in AD 405 and a model for linking public and personal integrity in *De Regimine*, a fifteenth-century English treatise on how to rule virtuously (Stocker 1998: 24–25).

During the Reformation and Counter-Reformation, Judith was an icon, an emblem, and inspiration for both Protestant and Catholic forces alike. Protestant portrayals in political treatises, dramatic productions, poetry, art, and even everyday tableware emphasised that the story of Judith was one of righteous resistance and showed that one could be civically virtuous in committing tyrannicide (Stocker 1998: 56–58). As Stocker (1998: 56) notes, 'she was an emblem of lay godliness who also implied divinely sanctioned revolt'. Not wanting its political power to be monopolised by their rival denomination, Catholics also appropriated the story of Judith for their cause, going as far as to organise theatrical productions of the tale in the vernacular. She also became closely associated with the Jesuits and the political theories of Juan de Mariana who advocated sedition and murder in defence of the Catholic faith (Stocker 1998: 59–61). Later, Judith was an icon for the French revolution and in its aftermath a symbol for those who opposed the reign of terror. She was eventually embodied in the figure of Charlotte de Corday, who self-identified with the heroine, after her notorious killing of Marat, and was later similarly depicted as an heir to the legacy of Judith on stage and on canvas (Stocker 1998: 111–119). More recently, Stocker (1998: 198–203) notes that Judith was revived as an important symbol for extreme forms of Zionism as an antecedent justification for ends–means calculations in the defence of the Israeli state and retribution against its enemies.

Stocker (1998: 173–197) argues that within the Anglo-German world, as the new sciences of eugenics, criminology, and psychiatry became established in the nineteenth century, portrayals of Judith increasingly became associated with both anti-Semitism and new forms of misogyny enabled by these forms of power-knowledge. These understandings of Judith found their antecedents in Renaissance forms of erotic art. By the later stages of the Victorian age, there is a raw sexuality that exudes from Judith with intimations that it may have been more than bloodlust that was satiated in Holofernes' tent. She is not demure or an individual overcome by circumstances. Rather, in being both dangerous and very desirable, Judith is the *femme fatale*, or in Freudian inspired psycho-analytic readings, a hysterical female consumed by penis envy.

The problem then over the centuries has been, how should Judith be read, represented, and understood? In her comprehensive genealogy of Judith, Stocker (1998) is able to provide multiple examples of how the story of Judith can be – and has been – understood in some contexts very negatively. Thus, as much as Judith has been elevated as a paragon of virtue, she has always been subject to projects of reclamation, redemption, and even removal. The views of Thomas Aquinas are quite instructive here. For Aquinas, the problem with Judith was how to weigh the righteousness of her deed and her own virtues with the deceitful means that she used to kill Holofernes (Carter 2005: 1). For Day (2001: 89–90), Judith's deployment of deceit is so widespread that it is either 'intrinsic to her nature or...she cannot see the difference between misleading one's enemy and misleading one's friends and tribe'. Thus, Stocker (1998: 24) has argued that over the centuries, Judith became read as the 'Good Bad Woman'. But there have also been counter-representations, such as that forwarded by Leopold von Sacher-Masoch in his infamous *Venus in Furs*, which reverse this portrayal in order to emphasise Judith as the '*Bad* Good Woman', moves that reposition her as a taboo object of sexual desire.

Judith, ambivalence, and the narration of assassination events

The ambiguity of Judith and the lingering ambivalence over this ambiguity is instructive. It provides additional insight on the recourse to targeted killing by liberal regimes as a means of security provision. The difficulty is the stickiness of assassination in general, and the story of Judith in particular, as emplotted actions through which targeted killing events are made meaningful. While the mechanisms and rationalities may echo the imperatives of biopolitics, there is the resonance of a moral problematique that lingers from the story of Judith, a problematique of balance that plays out in two interconnected ways. On the one hand, it is a problem that vexed Thomas Aquinas: how to establish a just balance between retribution and the means used to achieve it. In other words, what are we allowed to do, how must it be performed, whom are we justified doing it to, and under what conditions of insecurity do these actions become legitimate? Do we have more latitude in situations where we are managing the conduct of an outsider, a threat, or – in a psycho-analytic reading – one whose lascivious appetites make him/her, from a biopolitical perspective, ungovernable?

Initial answers to the these questions can be inferred from two recent high profile targeted killing events: the 2010 elimination of Mahmoud al-Mabhouh in Dubai, allegedly by Mossad agents, and the execution of Osama Bin Laden by US special forces in 2011. These two cases demonstrate the anxiety underpinning targeted killing events by regimes which identify themselves as liberal. On the one hand, the intended targets are to be dispassionately presented through a biopolitical register as threats to which extermination is the best response. On the other hand, there is also lingering anxiety about what it means to be an actor who commits this form of political violence: individualised, named killing. In particular, the perception of retributive motivations that are seen as central to cultural representations of assassination events pose significant problems, as do revelations of deceit.

The killing of Mahmoud al-Mabhouh

On 19 January 2010 Mahmoud al-Mabhouh was killed in his hotel room while visiting Dubai. This killing was alleged to have been undertaken by the Caesarea, an elite extermination unit within the Mossad. Mabhouh had already been subject to previous attempts on his life. In February 2009, he had survived a drone strike as part of a convoy in Sudan. Later that year, he narrowly missed being poisoned to death by a hit team in Dubai who were suspected of smearing noxious agents over fixtures in his hotel room (Spiegel Staff 2011). Although he had over a long career become a prominent arms procurer and intermediary for Hamas, it was claimed that these roles were secondary considerations for the state of Israel. It has been suggested in the media that Mabhouh had earned a 'Red Page' designation in 1989 for his direct involvement in the kidnapping and killing of two Israeli soldiers, Avi Sasportas and Ilan Saadon. As reported by Bergman (2011), and *Der Spiegel* (Spiegel Staff 2011), 'Red Page' is allegedly a code word used by the Mossad to denote individuals who have been marked for execution under the joint order of the Israeli Prime Minister and Minister of Defence. 'Red Pages' are not time sensitive and may take years to be fulfilled; they only expire if explicitly cancelled. His participation in these killings had returned to prominence within the Israeli intelligence services after Mabhouh gave an ill-advised interview to *Al Jazeera* in which he discussed the murders, the disposal of the bodies, his role, and his regrets about not having the opportunity to shoot the captives himself. Although his face was hidden in the televised report, Mabhouh was identified by his voice which had not been altered (Spiegel Staff 2011).

The operation was initially a complete success. Mabhouh was tracked by agents into Dubai, where they were able to determine his hotel, room number, and daily activities. After booking a room across the hall, the strike team carefully broke into his hotel room by manipulating the electronic door lock and awaited his arrival. Upon his return, members of the team promptly killed Mabhouh, placed him into bed, left the room, and locked the door behind them. As planned, his body was not discovered until all the field agents had left Dubai nearly a day later and his passing was originally treated as arising from natural causes.[7] An investigation into the event was only triggered when the Chief of Police for Dubai – Lieutenant General Dhahi Khalfan Tamim – received a phone call from Hamas alleging that Mabhouh had been killed by the Mossad. As he had travelled on a false passport under an assumed identity, Mabhouh's death had not registered as potentially sensitive when first discovered. Confirmation of the suspicious nature of the death was revealed through a subsequent analysis of surveillance camera footage that showed agents loitering in the hotel and observing Mabhouh's room (Bergman 2011).

Officially, Israel has neither confirmed nor denied the charges that it – or its agents – had any involvement in this targeted killing event. In public, Israeli officials were initially non-committal, but smug, about the operation. Katz

(2010) reported for the *Jerusalem Times* that upon leaving a cabinet meeting after news of the killing began to circulate amongst the world press:

> Interior Minister Eli Yishai told the reporters with a smirk on his face… 'All the security services make, thank God, great efforts to safeguard the security of the State of Israel' [while] Science and Technology Minister Daniel Herschkowitz added that 'My impression is that the Mossad knows how to get the job done, and it is a known thing that anyone who lifts a hand against a Jew is putting his life on the line.'

The success story soon turned into a global public relations disaster. Investigations by the police in Dubai exposed how Mossad agents had entered the state using foreign passports and stolen identities. These revelations strained diplomatic relations with the states in question and generated public outcry as stories began to circulate about how foreign nationals whose identities were stolen would need to clear their names with authorities in the United Arab Emirates. Similarly, with an eye to embarrassing the Israeli government, the Chief of Police in Dubai held regular press conferences in the weeks that followed. Their primary purpose was to publicly release CCTV footage of the Mossad agents involved in the operation, including clips of how they attempted to disguise themselves as tourists by wearing wigs and baseball caps while carrying tennis rackets (Spiegel Staff 2011). This footage, and how it was used by the Dubai police in conjunction with other techniques to uncover every aspect of the operation, further undermined the credibility of the Mossad and its director, Meir Dagen.

Amid the police investigation and subsequent media attention, what was taken as read was the rationale for the killing. Reports at the time stressed Mabhouh's position within the Hamas network, primarily as a deal-maker with close ties to Iran, as well as his commercial arms connections in China, Sudan, Dubai, and Syria. Media reports commented that the effort and risks involved for the Mossad appeared disproportionate to the disruptive potential for Hamas of eliminating Mabhouh. That retribution for his involvement in a specific act of violence was a significant motivating factor was only discovered in the months to follow. Ronen Bergman (2011) was one of the first to detail Mabhouh's 'Red Page' status and the way in which targeted killing operations by the Mossad were contextualised under the leadership of Meir Dagen. He reported that:

> Several Mossad operatives who have attended meetings in Dagan's office describe a ritual that he goes through when preparing a team for a dangerous mission. During the meeting, Dagan points to a large photograph hanging on his office wall of a bearded Jew wrapped in a prayer shawl, kneeling on the ground with his arms in the air. The man's fists are clenched, and his piercing eyes look straight ahead. Next to him stand two German SS officers, one holding a club and the other a pistol. 'This man,' Dagan

says, 'was my grandfather, Dov Ehrlich.' He then explains that shortly after the photo was taken, on October 5, 1942, his grandfather was murdered by the Nazis along with his family and thousands of other Jews in the small Polish town of Lukow.

'Look at this photograph,' Dagan tells the Caesarea fighters. 'This is what must guide us and lead us to act on behalf of the State of Israel. I look at the picture and vow that I will do everything I can to ensure that something like this will never happen again.'

(Bergman 2011)

Yet, revelations of the potential presence of retributive logics as well as the construction of existential threats by making historical analogies to Nazism and the Shoah have remained overshadowed by a fascination with the logistics of the killing. In particular, the diplomatic fallout over the use of passports and identities from the UK, Ireland, France, and Germany as well as forged American credit cards remained nodal points in media discussions of the event.

Similarly, reports from outside of Israel in the months that followed focused on what was described as the audacity of the Mossad in daring to conduct a targeted killing in a third country with whom relations were cool. This audacity was said to derive from the fact that the Mossad did not rate the competency of the authorities in Dubai (Bergman 2011). There was an assumption that agents would be in full control of the visibility of the operation and that they would remain unmarked in the visual field of their target and the local authorities. Thus, the goal was to take the form of an invisible presence in the aftermath, to leave traces but no definitive connections. However, the decision to kill Mabhouh also reflected a very specific answer to questions raised from the moral problematique of Judith: the tensions between retribution and the means used to achieve it. And in this instance, it was not so much that Mabhouh was killed – or how he was killed – but rather the logistical tactics used to facilitate the killing that ultimately brought moral sanction. Thus, the question of who can do what to whom was recast in this case. It was the means of opportunity (i.e., identity theft), and the nationalities of those who were affected by these means, that ultimately proved troublesome for the Israeli government.

The killing of Osama bin Laden

In the aftermath of the killing of Osama bin Laden, one could see again the lingering shadow of this same moral problematisation of 'who can do what to whom' in President Obama's official statement confirming the operation. The opening of his address uses predication and subject positioning to provide an initial justification for the act:

Tonight, I can report to the American people and to the world that the United States has conducted an operation that killed Osama bin Laden, the leader of Al-Qaeda, and a terrorist who's responsible for the murder of thousands of innocent men, women, and children.

(Obama 2011)[8]

Similar to Holofernes, bin Laden is positioned as a remorseless killer, who with the blood of thousands of innocents on his hands, moved himself beyond the possibility of being governed, politically engaged, or redeemed. Initial media reports – said to be from eye-witness accounts from unofficial sources – echoed these negative representational practices with claims that bin Laden used a woman as a human shield to protect himself during the raid and later, that he had a large cache of pornography on his computer (e.g., Bronstein 2013; Shane 2011). By echoing longer-standing Orientalist depictions of Muslims males, the point was clear: bin Laden was an individual whose actions, beliefs, and appetites placed his actions beyond any justificatory discourses, religious or otherwise. He marked himself for death. But that these points had to be made suggested anxiety over how widely shared this view might be.

There is another problematique of balance underpinning this question of dessert and its importance to the narratives of targeted killing events. It is about the balance of characteristics, appetites, and actions that constitutes the identity of the Self. Again, this problematique resonates in President Obama's speech. After creating a temporal sequence that began with a description of the horrors of September 11 – a move that recalled specific traumas while seeking to erase any memory of what had transpired before 9/11 – military actions that constitute the 'war on terror', and the decision to prioritise bin Laden upon assuming office, President Obama articulated an additional justification for this specific act of killing through a performative of the American Self:

> ...we will never tolerate our security being threatened, nor stand idly by when our people have been killed. We will be relentless in defence of our citizens and our friends and allies. We will be true to the values that make us who we are. And on nights like this one, we can say to those families who have lost loved ones to Al-Qaeda's terror: Justice has been done.
>
> (Obama 2011)

With an impressive rhetorical flourish, security and justice became instantiated in the actions of the United States and Obama confirmed that its security apparatus assumed the form of a contemporary 'Judith'.

What is at stake in this appropriation of Judith is not just 'Judith', as an iconic character of Judeo-Christian mythology, but how a liberal regime – as a potential assassin – understands itself as a moral actor and biopolitical entity. Many commentators have astutely drawn our attention to the performatives of sexuality and gender underpinning not only the Book of Judith but also the ways that similar acts of assassination have been understood (Sawyer 2001b; Stocker 1998). Anxiety over gender coding has then fed into moral questions regarding whether it is possible to remain righteous while undertaking an act that might otherwise be understood as unjust by introducing the fear that by partaking in the act, one may become perceived as weak, feminine, or hysterical. President Obama's pains to articulate that the killing of bin Laden was a prudential security measure that also provided justice without constituting a war on Islam speaks to an anxiousness over being perceived as hysterical.

In contrast then to biopolitical accounts that would stress the certainties of biomedical rationales and/or notions of effectiveness in making the case for the exception, the story of Judith reveals that there is always a sense of anxiety regarding the legitimacy of targeted killing that must be managed. Beyond the status of the targets or the reasons why an individual is to be targeted, these actions are in part recognised as events by audiences because of these very questions of legitimacy that position them as emplotted actions within a narrative landscape shaped by older assassination events. It is not merely a question of how targeted killing provides security but also what its commissioning reveals about the perpetrator. Judith was able to claim that her hands were guided by God, a claim that others chose to accept, and a claim that fit within the dominant iterative of the day. What though is guiding the hands of liberal states that engage in targeted killing and do the potential contenders have the same moral force in the contemporary world as a deity did in the ancient? Or is this one of the myriad problems that the (re)turn to religious fundamentalism – i.e., the 'will to revenge' identified by Connolly (2005) – in liberal political spheres attempts to solve?

At the heart of the contemporary debate about targeted killing and the ways that the Book of Judith has contributed to the constitution of a (contested) configurational system, mode of historical explanation, and forms of normative intervention, is an ambivalence about liberalism: is liberalism a good system that can produce bad outcomes or a bad system that can produce good outcomes? This is a question bubbling beneath the surface of the biopolitical age. And it is profoundly unsettling for those who understand themselves to be liberal agents, subject to a benevolent governing regime.

Conclusions

Orna Ben-Naftali and Keren R. Michaeli (2003: 369) remind us that 'targeted killings do not take place in a vacuum. They reflect a deliberate administrative decision...'. As administrative decisions, targeted killings are embedded into the social, political, cultural, and economic modalities through which they are contemplated. Modalities are themselves constituted and shaped by relations of power. Power-relations, their flows, their dynamics, and the ways in which these link, connect, and network are never uniform or even internally consistent. Assassination and targeted killing embody cynegetic relations of power that separate predator from prey. They also establish scripts through which violence is enacted and understood. The cultural resonance of the 'will to revenge' and ambivalence over its legitimacy are important elements in the presentation of assassination and targeted killing as recognisable events – i.e., instances of political violence intended to achieve security objectives – as well as the processes that make them possible. While the analysis of security politics tends to focus on the use of the exception and its biopolitical underpinnings, the story of Judith shows emplotted actions are more complex and potentially contradictory than the technical rationalisations that contribute to them. It is therefore important to examine how the emplotted actions of political violence are produced and circulated through cultural practices.

Stocker (1998: 61) argues that if there is one constant in the assassination problematic from an historical perspective, it is that 'Judiths were unruly creatures, to be exploited when one's own objectives were disruptive, but discomforting once one had achieved power oneself'. Israel and the United States by figuratively assuming the forms of Judith to embody 'the will to revenge' through their increased use of targeted killing would thus appear to be somewhat anomalous. By engaging in a mimetic appropriation of Judith, these regimes lock themselves into a narrative imaginary that both believes in the ability of an ancient tactic to manage today's strategic landscape and the righteousness of its commissioning against specific types of people. Yet, inherent to playing the role of Judith are performative risks. It is this lingering ambivalence over what the Self becomes through their commission – and how the Self is perceived by others – that may help to counter-act its biopolitical underpinnings and produce a renewed reluctance to engage in the practice.

The ambivalence of assassination and targeted killing also cultivates probing and politically substantive questions about targets that might otherwise be evaded through the invocation of biopolitical rationalisations. Of particular importance is how the instability of danger becomes concretised in security imaginaries that then make this form of violence against particular people appear permissible. Thus, what the proceeding analysis has shown – in part – is that there is an interesting space for contestation at the point where events emerge and become understood as 'assassination' or 'targeted killing' events. However, as the contestation of meaning over the representation of Judith attests, the role of ambivalence is not necessarily direct or necessarily uni-directional. It is constituted by divergent forms of resonance and assemblages that take shape, transform, and even dissipate over time and space.

The cases explored in this chapter are somewhat unique in contemporary counter-insurgency in terms of their high profile. Increasingly, while kill/capture raids of the type described in this chapter continue to be the predominant mode of targeted killing, public attention has been drawn away from close quarters eliminations towards targeted killing events conducted from remotely piloted aircraft (RPAs) (Masters 2013). Yet, drone strikes are still locked into the orbit of a cultural narrative that understands these events as potentially involving treachery, dishonourable conduct, and unseemly activity. Even though targeted killing is potentially permissible according to the procedural requirements of the laws of war, the insertion of the drone as the means is still not able to escape the gravitation pull of this lingering anxiety. Thus, in addition to the legal and biopolitical arguments that can be made in support of targeted killing, the assemblage also draws upon broader cultural values that emerge from contemporary political economy in advocating for the development and deployment of remotely piloted technologies. A discussion of these values and their political economy is the focus of the next chapter.

Notes

1 Species here should be understood more broadly than traditional textbook understandings. It includes genetics as well as phenomes with respect to political, economic, and social behaviours that emerge from the interplay of genotypes in a given environment. Phenotypic characteristics are particularly important as biopolitical forms of governance often use them as a way of reverse engineering into the species genotype.
2 A recently leaked Pentagon study reported that signals intelligence (primarily National Security Agency surveillance) provided approximately two-thirds of the data used to find high value individuals for targeted killing while full motion video (from drones) was the primary method of 'fixing' (i.e., confirming the location) and 'finishing' (i.e., attempting to kill) targets. See ISR Task Force (2013).
3 There are those who dispute these findings such as Johnston (2012) while others argue that effectiveness is highly dependent on a series of contextual factors (e.g., CIA 2009; Freeman 2014; Jordan 2014b).
4 Important here is to note that assassination and targeted killing are often interchangeable in common linguistic use and meaning (i.e., as individualised killing committed by the state and/or its agents), despite technical differences in how they are defined under the law.
5 Passages of scripture are taken from the King James Bible edited by Carroll and Prickett (1997).
6 Janes (2005: 3) argues that '...decapitation as violence...flourishes with culture, [and was] dependent on technological advances'. For a detailed genealogy of decapitation as both symbolic practice and form of violence, see Janes (2005).
7 The death was eventually recorded as being caused by a combination of poison then smothering, though it is now claimed that the actual cause of death remains unknown (see Bergman 2011).
8 For a discussion of predication and subject positioning in discourse analysis, see Doty (1996: 10–12).

Bibliography

Aarons, K. (2013) 'Cartographies of Capture', *Theory & Event*, 16(2).
Agamben, G. (1998) *Homo Sacer: Sovereign Power and Bare Life*, Stanford, Stanford University Press.
Beck, U. (1992) *Risk Society: Towards a New Modernity*, London, Sage Publishing.
Benjamin, W. (1986) 'A Critique of Violence', in P. Demetz (ed.), *Reflections: Essays, Aphorisms, Autobiographical Writings*, New York: Schocken Books, 277–300.
Ben-Naftali, O. and Michaeli, K. R. (2003) 'Justice-Ability: A Critique of the Alleged Non-Justiciability of Israel's Policy of Targeted Killings', *Journal of International Criminal Justice*, 1(2), 368–405.
Ben-Yehuda, N. (1990) 'Gathering Dark Secrets, Hidden and Dirty Information: Some Methodological Notes on Studying Political Assassinations', *Qualitative Sociology*, 13(4), 345–371.

Ben-Yehuda, N. (1997) 'Political Assassination Events as a Cross-cultural Form of Alternative Justice', *International Journal of Comparative Sociology*, 38(1–2), 25–47.

Bergman, R. (2011) 'The Dubai Job'. *Gentleman's Quarterly (GQ)*. Available from: http://www.gq.com/news-politics/big-issues/201101/the-dubai-job-mossad-assassination-hamas. Accessed 15 October 2015.

Bhabha, H. (2005 [1994]) *The Location of Culture*, Abingdon, Routledge.

Bronstein, P. (2013) 'The Man Who Killed Osama bin Laden… Is Screwed'. *Esquire*. Available from: http://www.esquire.com/news-politics/a26351/man-who-shot-osama-bin-laden-0313. Accessed 15 October 2015.

Bunker, R. J. and Begert, M. (2002) 'Operational Combat Analysis of the Al Qaeda Network', *Low Intensity Conflict and Law Enforcement*, 11(2), 316–339.

Campbell, D. (1998) *Writing Security: United States Foreign Policy and the Politics of Identity*, Minneapolis, University of Minnesota Press.

Carroll, R. and Prickett, S. (eds) (1997) *The Bible: Authorized King James Version*, Oxford, Oxford University Press.

Carter, M. (2005) 'Judith: Headhunting for Virtue'. *Damascus Gate*. Available from: http://www.damascusgate.org/Judith_Headhunting.pdf. Accessed 15 October 2015.

Carvin, S. and Williams, M. J. (2014) *Law, Science, Liberalism, and the American Way of Warfare*, Cambridge, Cambridge University Press.

Central Intelligence Agency (2009) 'Best Practices in Counterinsurgency: Making High Value Targeting an Effective Counterinsurgency Tool'. Central Intelligence Agency. Available from: https://wikileaks.org/cia-hvt-counterinsurgency. Accessed 15 October 2015.

Chamayou, G. (2012) *Manhunts: A Philosophical History*, Princeton, Princeton University Press.

Connolly, W. E. (2005) 'The Evangelical-Capitalist Resonance Machine', *Political Theory*, 33(6), 869–886.

Connolly, W. E. (2011) 'The Politics of the Event'. *The Contemporary Condition*. Available at http://bit.ly/lMlbQ0. Accessed 15 October 2015.

Coward, M. (2009) 'Network-Centric Violence, Critical Infrastructure and the Urbanization of Security', *Security Dialogue*, 40(4–5), 399–418.

Crawford, G. A. (2009) 'Manhunting: Counter-Network Organization for Irregular Warfare', *DTIC Document*. Available from: http://oai.dtic.mil/oai/oai?verb=getRecord&metadataPrefix=html&identifier=ADA514554. Accessed 15 October 2015.

Croft, S. and Moore, C. (2010) 'The Evolution of Threat Narratives in the Age of Terror: Understanding Terrorist Threats in Britain', *International Affairs*, 86(4), 821–835.

David, S. R. (2003a) 'If Not Combatants, Certainly Not Civilians', *Ethics & International Affairs*, 17(1), 138–140.

David, S. R. (2003b) 'Israel's Policy of Targeted Killing', *Ethics & International Affairs*, 17(1), 111–126.

Day, L. (2001) 'Faith, Character, and Perspective in Judith', *Journal for the Study of the Old Testament*, 95(26), 71–93.

Dean, M. (1999) *Governmentality: Power and Rule in Modern Society*, London, Sage.

De Goede, M. (2008) 'Beyond Risk: Premediation and the Post-9/11 Security Imagination', *Security Dialogue*, 39(2–3), 155–176.

Deleuze, G. (1992) 'Postscript on the Societies of Control', *October*, 59(Winter), 3–7.

deSilva, D. A. (2006) 'Judith the Heroine? Lies, Seduction, and Murder in Cultural Perspective', *Biblical Theology Bulletin: A Journal of Bible and Theology*, 36(2), 55–61.

Dillon, M. (2007) 'Governing through Contingency: The Security of Biopolitical Governance', *Political Geography*, 26(1), 41–47.

Dillon, M. and Reid, J. (2009) *The Liberal Way of Warfare: Killing to Make Life Live*, London, Routledge.

Doty, R. L. (1996) *Imperial Encounters: The Politics of Representation in North-South Relations*, Minneapolis, University of Minnesota Press.

Eisenberg, Uri D.Dan, and Eli Landau (1978) *The Mossad: Israel's Secret Intelligence Service Inside Stories*, London, Paddington Books.

Esler, P. F. (2002) 'Ludic History in the Book of Judith: The Reinvention of Israelite Identity?', *Biblical Interpretation*, 10(2), 107–143.

Feldman, A. (1991) *Formations of Violence: The Narrative of the Body and Political Terror in Northern Ireland*, Chicago, University of Chicago Press.

Ford, F. L. (1985) *Political Murder: From Tyrannicide to Terrorism*, Cambridge, Harvard University Press.

Foucault, M. (1990) *The History of Sexuality: An Introduction*, New York, Random House.

Foucault, M. (2007) *Security, Territory, Population: Lectures at the College de France 1977–1978*, Basingstoke, Palgrave.

Freeman, M. (2014) 'A Theory of Terrorist Leadership (and its Consequences for Leadership Targeting)', *Terrorism and Political Violence*, 26(4), 666–687.

Garland, D. (2005) 'Penal Excess and Surplus Meaning: Public Torture Lynchings in Twentieth-Century America', *Law & Society Review*, 39(4), 793–833.

Gregory, D. (2004) *The Colonial Present*, Oxford, Blackwell Publishing.

Hafez, M. M. and Hatfield, J. M. (2006) 'Do Targeted Assassinations Work? A Multivariate Analysis of Israel's Controversial Tactic during Al-Aqsa Uprising', *Studies in Conflict and Terrorism*, 29(4), 359–382.

Hansen, L. (2006) *Security as Discourse: Discourse Analysis and the Bosnian War*, Abingdon, Routledge.

Honig, O. (2007) 'Explaining Israel's Misuse of Strategic Assassinations', *Studies in Conflict and Terrorism*, 30(6), 563–577.

Huysmans, J. (2008) 'The Jargon of Exception – On Schmitt, Agamben and the Absence of Political Society', *International Political Sociology*, 2(2), 165–183.

ISR Task Force, (2013) 'ISR Support to Small Footprint – CT Operations in Somalia/Yemen'. *Pentagon*. Available from: https://theintercept.com/document/2015/10/14/small-footprint-operations-2-13. Accessed 15 October 2015.

Jabri, V. (2006) 'War, Security and the Liberal State', *Security Dialogue*, 37(1), 47–64.

Janes, R. (2005) *Losing Our Heads: Beheadings in Literature and Culture*, New York, NYU Press.

Johnston, P. B. (2012) 'Does Decapitation Work? Assessing the Effectiveness of Leadership Targeting in Counterinsurgency Campaigns', *International Security*, 36(4), 47–79.

Jones, C. A. (2015) 'Frames of Law: Targeting Advice and Operational Law in the Israeli Military', *Environment and Planning D: Society and Space*, 33(4), 676–696.

Jordan, J. (2009) 'When Heads Roll: Assessing the Effectiveness of Leadership Decapitation', *Security Studies*, 18(4), 719–755.

Jordan, J. (2014) 'Attacking the Leader, Missing the Mark: Why Terrorist Groups Survive Decapitation Strikes', *International Security*, 38(4), 7–38.

Jordan, J. (2014b) 'The Effectiveness of the Drone Campaign against Al Qaeda Central: A Case Study', *Journal of Strategic Studies*, 37(1), 4–29.

Katz, Y. (2010) 'Security and Defense: Espionage, with a Smile'. *The Jerusalem Post Online.* Available from: http://www.jpost.com/Features/Front-Lines/Security-and-Defense-Espionage-with-a-smile. Accessed 15 October 2015.

Kessler, O. and Werner, W. (2008) 'Extrajudicial Killing as Risk Management', *Security Dialogue*, 39(2–3), 289–308.

Kober, A. (2007) 'Targeted Killing during the Second Intifada: The Quest for Effectiveness', *Journal of Conflict Studies*, 27(1), 76–93.

Massumi, B. (2007) 'Potential Politics and the Primacy of Preemption', *Theory and Event*, 10(2).

Masters, J. (2013) 'Targeted Killings', *Council on Foreign Relations.* Available from: http://www.cfr.org/counterterrorism/targeted-killings/p9627. Accessed 15 October 2015.

Mayfield, T. (2009) 'The Accounts of Deborah (Judges 4–5) in Recent Research', *Currents in Biblical Research*, 7(3), 306–335.

Mills, C. W. (2000) *The Sociological Imagination*, Oxford, Oxford University Press.

Morehouse, M. (2014) 'It's Easier to Decapitate a Snake than It Is a Hydra: An Analysis of Colombia's Targeted Killing Program', *Studies in Conflict & Terrorism*, 37(7), 541–566.

Muller, B. J. (2008) 'Securing the Political Imagination: Popular Culture, the Security Dispositif and the Biometric State', *Security Dialogue*, 39(2–3), 199–220.

Obama, B. (2011) 'Remarks by the President on Osama bin Laden', *The White House.* Available from: http://www.whitehouse.gov/blog/2011/05/02/osama-bin-laden-dead. Accessed 10 June 2011.

Pan, D. (2009) 'Against Biopolitics: Walter Benjamin, Carl Schmitt, and Giorgio Agamben on Political Sovereignty and Symbolic Order', *The German Quarterly*, 82 (1), 42–62.

Patterson, E. and Casale, T. (2005) 'Targeting Terror: The Ethical and Practical Implications of Targeted Killing', *International Journal of Intelligence and Counter-Intelligence*, 18(4), 638–652.

Reis, P. T. (2005) 'Uncovering Jael and Sisera: A New Reading', *Scandinavian Journal of the Old Testament*, 19(1), 24–47.

Ricoeur, P. (1973) 'The Model of the Text: Meaningful Action Considered as a Text', *New Literary History*, 5(1), 91–117.

Said, E. (1978) *Orientalism*, New York, Pantheon.

Sawyer, D. F. (2001a) 'Dressing Up/Dressing Down: Power, Performance, and Identity in the Book of Judith', *Theology and Sexuality*, 15(8), 23–31.

Sawyer, D. F. (2001b) 'Gender Strategies in Antiquity: Judith's Performance', *Feminist Theology*, 10(28), 9–26.

Shane, S. (2011) 'Pornography Is Found in Bin Laden Compound Files, US Officials Say'. *New York Times Online*, Available from: http://www.nytimes.com/2011/05/14/world/asia/14binladen.html?_r=0. Accessed 15 October 2015.

Slotkin, R. (1973) *Regeneration through Violence: The Mythology of the American Frontier, 1600–1860*, Norman, University of Oklahoma Press.

Slotkin, R. (1992) *Gunfighter Nation: The Myth of the Frontier in Twentieth-century America*, Norman, University of Oklahoma Press.

Spiegel Staff. (2011) 'An Eye for an Eye: The Anatomy of Mossad's Dubai Operation'. *Der Spiegel Online.* Available from: http://www.spiegel.de/international/world/an-eye-for-an-eye-the-anatomy-of-mossad-s-dubai-operation-a-739908.html. Accessed 15 October 2015.

Stocker, M. (1998) *Judith Sexual Warrior: Women and Power in Western Culture*, New Haven and London, Yale University Press.

Wallhead, C. (2001) 'The Story of Jael and Sisera in Five Nineteenth and Twentieth Century Fictional Texts', *Atlantis*, 23(2), 147–166.

Weber, M. (2005) 'Bureaucracy', in H. H. Gerth and C. W. Mills (eds), *From Max Weber: Essays in Sociology*. 2, London: Routledge, 196–244.

Weldes, Mark J.Laffey, HughGusterson, and Raymond Duvall (1999) 'Introduction: Constructing Insecurity', in*Cultures of Insecurity: States, Communities, and the Production of Danger*, Minneapolis: University of Minnesota Press, 1–33.

Williams, M. J. (2008) '(In)Security Studies, Reflexive Modernization, and the Risk Society', *Cooperation and Conflict*, 43(1), 57–79.

Ziolkowski, T. (2009) 'Re-Visions, Fictionalizations, and Postfigurations: The Myth of Judith in the Twentieth Century', *Modern Language Review*, 104(2), 311–332.

4 Science, capitalism, and the RPA

Introduction

As suggested in the previous chapter, liberal regimes like Israel, the United Kingdom, and the United States are caught between a perceived need to engage in targeted killings and an ambivalent cultural landscape where targeted killing events can be placed within the grammatical structures of assassination. In the current wars of the 'colonial present' (see Gregory 2004), waged overtly in Afghanistan, Iraq, Libya, and Syria and covertly in Pakistan, Somalia, and Yemen, it has been the Remotely Piloted Aircraft (RPA) – or drone – that has prominently featured as the technological expression of the targeted killing assemblage. Surveying, monitoring, and observing the battle-space from afar, drones are represented as the pinnacle of modern military technology. The RPA is central to waging 'low risk war' in which it is claimed that casualties can be limited to the enemy through network analyses, precision strikes, and the absence of a theatre-based human operator. But given that RPAs are often engaged in targeted killings, actions whose *prima faci* legality under the laws of war are circumscribed by the gravitational pull of cultural understandings of assassination events, what does their prominent position in contemporary discourses of counter-insurgency reveal about the politico-cultural dynamics that underpin targeted killing? And what is it about the RPA that has allowed it not only to be given a special status in popular military discourses but to have become privileged within military circles themselves?[1] How has it become possible for advanced militaries to become addicted to drones (Smith 2010)?

There are potential strategic, operational, and tactical advantages provided by the deployment of the RPA. At the same time, some of the advantages conferred by the RPA are not independent of the network centric and chaoplexic war-fighting doctrines that have become predominant in military thinking over the past three decades. While initially conceived as part of wider processes of military transformation, the concepts of netwar and chaoplexity have shaped more recent forays into the irregular warfare said to characterise 'overseas contingency operations' (Bousquet 2009; Guha 2011). Thus, the current context in which the RPA has become a central part of advanced war-fighting is comprised of a complex constellation of theorisations and practices underpinning the dynamics of contemporary conflict in which how to best conduct war is problematised in relation to scientific principles with socio-economic resonance. But how have these recent understandings – and the practices that flow from them – become possible?

To answer this question, Antoine Bousquet (2009) has traced how understandings of war and science have been mutually constitutive, at least since the development of biological mechanism in the seventeenth century. His analysis therefore provides a useful starting point for understanding the intersections of technology, discourse, and war-fighting. In this chapter, I argue that in RPA discourse, the privileges given to notions of speed, automation, surveillance, information, de-skilling, and spectacle in the production of the 'kill chain' indicates that broader influences are making a contribution (USAF 2007). Specifically, cultural values found in twenty-first-century capitalism are shaping an integral aspect of counter-insurgency warfare today. Thus, capitalist values are intimately involved in processes of coding and recoding as well as contributing to the variable problematisations, processes of accumulation, and legitimation discourses that territorialise the targeted killing assemblage.

To demonstrate these relations, this chapter is divided into three main sections. In the first, the important contribution made by Bousquet to the understanding of war is outlined. While his analysis provides an excellent overview of how science, technology, and theory have influenced the practices of war, I offer one modification to his argument: that the influence of values featured in commercial discourses on both scientific discovery and combat must be taken into account if one wishes to understand the 'complex social assemblage' constituted by war. In the second section, I provide a general means of conceptualising the discursive dynamic of science, war, and capitalism at the present juncture. In the third section, I demonstrate how elements that are identified as characteristics of contemporary capitalism resonate in discussions of the current utility and future development of the RPA.

The scientific way of warfare

In *The Scientific Way of Warfare: Order and Chaos on the Battlefields of Modernity*, Bousquet (2009) provides a rich genealogical analysis of understandings of war. He demonstrates that war doctrines traditionally advanced forms of hierarchical command and centralisation. Over time, these transformed into doctrines that forwarded the principles of decentralisation and the horizontal integrating functions of networks in contemporary war-fighting. His central claim is that warfare, as a profoundly social dynamic, is inextricably linked to science. Bousquet argues that as a methodological framework for problem-solving and as a way of defining problems as such, science is the dominant paradigm of understanding in modernity. In other words, the presuppositions used to acquire knowledge of – and to construct schematics of – the natural, mechanical, and digital worlds influence the ways in which social phenomena are also understood (Bousquet 2009: 3; see also Guha 2011: 43–85).

War-fighting doctrines throughout modernity have been shaped by the 'dominant corpus of scientific ideas' of the context within which particular wars – or preparation for war – have taken place (Bousquet 2009: 3). This interlinking has not only involved the transposition of scientific thought into war-fighting doctrines but also the incorporation of new technologies developed at the cutting edge of scientific discovery into the 'complex social assemblages of war' (Bousquet 2009: 3). As Bousquet (2009: 3) suggests, '...scientific ideas have been systematically recruited to inform thinking about the very nature combat and the forms of military organisation best suited to prevail in it'. War and science have therefore been problematised in relation to one another. He refers to this interplay of scientific thought, technology, and war as the *scientific way of warfare*, 'an array of scientific rationalities, techniques, frameworks of interpretation, and intellectual dispositions which have characterised the approach to the application of socially organised violence in the modern era' (Bousquet 2009: 4).[2] In the current epoch, Bousquet identifies the paradigmatic regime of war as chaoplexic with the network serving as its organising concept. Chaoplexity owes allegiance to the concepts of non-linearity, positive feedback, self-organisation, and emergence – elements central to Dillon and Reid's (2009) understanding of the liberal way of war discussed in the introductory chapter. Chaoplexic theories have influenced military strategy and tactics to privilege decentralisation and the rapid adaptability of the 'swarm' (i.e., independent yet interconnected groupings capable of rapid adaptation to tactical conditions through the efficient dissemination and processing of shared information) in contemporary combat (Bousquet 2009: 30–35).

Bousquet's detailed account of the scientific way of warfare is based on a view of science that emphasises its ability to predict and explain phenomena in order to exert control and power over them. This view of science is congruent with what he identifies – via Martin van Creveld – as a paramount concern for military practice from time immemorial: obtaining certainty within the chaotic environment of combat. For Bousquet (2009: 10), warfare, like science, should therefore be understood as '...the attempt to impose order over chaos, to exert control where it most threatens to elude, and to find predictability in the midst of uncertainty'. While military action is rationalised in relation to broader (geo)political aims and may be shaped by moral, legal, and ethical codes of conduct, strategic theorisation seeks to '...bring order and predictability to activities which would otherwise be left entirely to chance and contingency' (Bousquet 2009: 10). Central to this ordering imperative have been technologies that served not only as the necessary equipment for reaching solutions to specific scientific problems that may have proceeded their invention, but also as an impetus for posing new problems that in turn led to the advancement of knowledge. New theories engendered by technology either improved these specific machines and the requisite forms of knowledge underpinning them or made possible new equipment and methods of knowledge production. All of these interconnections in turn fostered the expansion of novel scientific knowledge (Bousquet 2009: 16). But, as Bousquet (2009: 2) argues, '...technology is first and foremost a tool and one that only takes on meaning and purpose within the specific social and cultural formations in which it is deployed'. Thus, while he emphasises the scientific and technological dimensions in his analysis, Bousquet is not neglectful of the role of social organisation. He argues that technology and knowledge production are always embedded into a social organisation and structures of management (Bousquet 2009: 17–19). Within these complex formations, one cannot with any precision determine a principal catalyst or a direction of causality. Borrowing from the work of Gilles Deleuze and Felix Guattari, Bousquet illustrates how the interplay of material and discursive forces with the constellation formed by technology and the social environment should be seen themselves as 'machinic'. This means that these forces form a complex apparatus whose interconnected elements operate in unison (Bousquet 2009: 19). Thus, Bousquet (2009: 19) argues that one can see the concurrent operationalisation of '…technical apparatuses, social organisations, and military thinking'. Yet within these complex apparatuses, as Manabrata Guha (2011: 138–139) suggests, changes to the scientific way of warfare have not transformed the parameters how war is understood: its central problematisation – i.e., how to maintain control in an environment said to be characterised by chance and uncertainty – has remained constant; what has changed are the means by which it is thought possible to do so.

Although rich, detailed, and compelling, explicitly missing from Bousquet's account of science, warfare, and the scientific way of warfare is the accounting of a significant aspect of any social assemblage: the manner in which the assemblage under consideration has been organised economically. There are at least three inter-related dynamics that require consideration with respect to the forms of modern scientific warfare examined by Bousquet. First are the multi-directional flows – material and discursive – that might conjoin science and economic structures – and more specifically, during the modern period, science and markets. Second is how warfare itself may result from economic dynamics and tensions that arise from within particular systems of material accumulation. Thus, the why of war (i.e., why must one fight) and the how of war (i.e., the strategies, tactics, and technologies to be deployed) – important elements in war-fighting doctrines – could be related to how forms of accumulation produce particular conflict problematics (i.e., frameworks of understanding that present the conflict as a specific type of problem) and institutional structures. Third are the interrelationships between forms of military organisation and forms of economic organisation. These can be conceptualised in terms of structures and associated discourses such as political economy, logistics, and management. These also include ontological and epistemological assumptions. How these assumptions and discourses have been conceptualised across a range of literatures will be discussed in the following section. While such a discussion can only be indicative in the scope of the generalisations offered, its purpose is to show how science, war, and economic organisation have historically been intertwined and the diversity of these inter-relationships.

Science, war, and economic organisation

It is not uncommon for science to be positioned as an independent search for truth. The argument is that science is not – and should not be – influenced by ideological inclinations. For Karl Popper (2011), it was the very absence of ideology that was said to define science. Yet, the work of philosophers such as Paul Feyerabend (1999) and historians like Thomas Kuhn (2012) has under-mined the narrative of detachment within the practices of science and the production of scientific knowledge. The power-relations that shape scientific knowledge production are not intrinsic to science itself or to the pathologies of the academy. Sociological analyses of the development of modern science contend that there are clear connections that link scientific disciplines to eco-nomic dynamics. Andrew Ross (1996: 5–6) argues that since Boris Hessen's 1931 lecture on the ties between Newton's *Principia* and the needs of a nascent English bourgeoisie, the 'indebtedness of empirical science to market interests' has been clear. Robert K. Merton (1973: 36) was more circumspect about the catalytic 'role of needs in determining the thematics of scientific research.' Yet, he was still interested in exploring the differentiating impacts of social forces and economic interests on the scientific revolutions of the seventeenth century. Toby E. Huff (2003: 14) has argued that within the literature on the development of modern science and the development of capitalism, the rela-tionships between economic organisation and science are often presented as mirror images. He demonstrates how several historians of scientific development including Max Weber and Joseph Needham argued that the rise of modern science in the West – and the relative lack of parallel developments in other regions – reflected the localised development of capitalism itself in the Euro-Atlantic corridor. For Weber and others, it was characteristics inherent to Protestantism that facilitated scientific advancement in the West, despite the lengthy head-starts in astronomy, mathematics, and even chemistry experienced outside of these particular societies.

The converse is also relevant. The ways in which economic dynamics are structured, managed, and understood are often dependent on scientific dis-courses. Scientific discourses may shape questions that are asked and where answers are sought. While the culturally deterministic aspects of prior arguments about the dynamics formed by science and the economy are problematic, science itself – both directly and indirectly through the transposition of its ideas, dis-courses, and technologies – has played an important role in the development, regulation, and deregulation of market economies and their practices. The argument raised here is not to suggest that science has necessarily become tainted as a result of these dynamics. The point is that if we are to understand war, it is important to locate science and war within the socio-economic dynamics in which both are being problematised and practiced.

Similarly, the relationship between forms of accumulation and warfare has been long recognised. Structural realists, including Paul Kennedy (2010), have charted how the rise and fall of great powers involves the military balance of power in a given era as well as the economic development and industrial resiliency of these states. Marxian thinkers including Lenin (1999), Kautsky (2007), and Hobson (1938) linked imperial warfare to the contradictions inherent to the functioning of advanced capitalist economies. Variations of the theme of imperial war were later picked up in world systems analysis as well as strands of dependency theory that argued the exploitation of the developing world was enabled by collusion between elites of the first and third worlds. The maintenance of these systems depended on the suppression of resistance, through violent means as required. Similarly, with the ending of World War II, concerns began to be raised with regards to how industries reconfigured for the war effort remained focused on the production of armaments and military technology. More important, it was argued that rather than abandoning the arms industry and refocusing production towards civilian needs, these industries had organised into a powerful lobby to promote military research, development, and manufacturing in order to maintain profitability (Baran and Sweezy 1968). Known as the 'military-industrial complex', analysts like Seymour Melman (1970) and C. W. Mills (1956) argued that this 'power elite' was comprised of a shifting (and often revolving) conglomeration of industrialists, lobbyists, politicians, military officials, and members of the media who actively exaggerated threats facing the United States while promoting the need for specific weapons systems, and the requisitely high levels of military spending to support these systems. Culturally, as a social force, the military industrial complex could be argued to have won the war of position in Cold War America and contributed both directly and indirectly to a culture of fear (Campbell 1998; Robin 2004). Outside of core capitalist states, multiple analyses of the so-called 'New Wars' of the post-Cold War era have argued that these conflicts and the social formations that emerge within them (e.g., warlordism) are directly influenced by market forces (Reno 1998; Duffield 2001). In response, Michael Hardt and Antonio Negri (2000) argued that contemporary war-fighting, often framed in terms of humanitarian interventions that aim to mitigate the deleterious effects of disorder in the periphery, reflects the global de-territorialisation of governance and commerce.

But the intersections of military and economic structures go beyond how economic considerations may contribute to conflict or the means by which war is waged. For example, rich historical analyses of the Renaissance period – often with a focus on the Mediterranean region – have explored the intersections of war, science, art, and economics. Transformations in siege warfare brought about by advancements in artillery were fostered by developments in the European knowledge of geometry. Geometrical advancement was often motivated by the pursuit of more accurate systems of navigation and cartography. The processes of mapping made possible from improvements in geometrical understanding and artistic expression integrated elements of 'textual analysis, computation, and visualisation to which scholars, artists, merchants, and patrons contributed working side by side' (Fiorani 2005: 5). The results of these endeavours gave space for new forms of artistic expression. Moreover, these discoveries provided the surveying, mapping, navigating, and ballistic calculations that would make contributions to the growth and spread of markets, the emerging political economy of European colonialism, and the development of the modern state into a functional unit of territorial governance.

Similarly, the impact of military organisation on the development of nineteenth-century European societies and capitalism was a focus of early political sociology. Richard Sennett (2006) argues that Max Weber's work on the militarisation of civil society in Prussia – and subsequent work by Joseph Schumpeter – illustrates a very important dynamic that contributed to the sustainability of capitalism. As a form of market structure, capitalism had initially been poorly organised at the level of the firm. Disorganisation at the local level made firms – and the system itself – susceptible to spectacular collapse. Sennett (2006: 30) argues via Weber that the military model of organisational management that included clear chains of command and duties, which were logically delineated, was implemented at the level of the firm as a means of providing effective management. Inclusion, albeit within a strict hierarchy, was presented as the most desirable primary organisational logic as opposed to efficiency. A small group of individuals were to give orders while a larger group of subordinates were to follow.[3] In addition, a more disciplined labour force was produced, often through drill and forms of internal monitoring. This shift from primarily external to a mixture of external and internal forms of monitoring was outlined in the work of Michel Foucault. In *Discipline and Punish*, Foucault (1977) revealed the interplay of military logics and the emerging human sciences in the contemporaneous rise of the prison, public education, and public health as prominent institutions in the nineteenth century.

But as Sennett (2006) outlines, the cultural power of militarism extended beyond the factory or mill as investment itself began to be framed in terms of military concepts like campaigns, strategic thinking, and outcome analyses. Thus, Sennett (2006: 23) suggests that the problems within capitalism that were to be addressed by these measures were framed as problems of disorder, unpredictability, and temporality. This framing was very similar to the inherent problems Bousquet (2009) identifies as central to scientific warfare: how to obtain and maintain predictability and control over events rapidly transpiring within a complex environment. Thus planning was undertaken with a long-term perspective privileged within decision making. In addition, the aversion to the unpredictability of change contributed to forms of organisational conservatism, with any modifications to standard operating procedures being slow-paced and incremental.

While the approaches, views, and judgements about science, war, and economic organisation have differed, the discussion above has shown how these three elements have been previously conceptualised. The treatment has been largely artificial in that the dynamics described above have been identified in isolation to one another. Moreover, this discussion has largely focused on analyses of the historical past as opposed to the present. But what is instructive is that inter-relationships between science, war, and economics have features across a range of scholarship. In the following section, the triad of science, war, and economic organisation will be further specified with regards to the discourses that constitute the contemporary culture of capitalism.

Capitalism, science, and war

How then does the current scientific way of warfare intersect with the complex social assemblage in which counter-insurgency and war-fighting are currently problematised? In providing a conceptual model of the forms that these connections may take and the relations of power that shape them, one is confronted with several difficulties. The first is that '[scientific development] can be both a cause and an effect, that [is], the social and economic are both causes and effects [and] all are "emergent and structuring"' (Arnold 2003: 239). The second difficulty is capitalism itself is not an easily delineated analytic category.[4] What is being assumed for the purposes of the analysis that will follow is that capitalism is as much about the instantiation of specific cultural values, styles of thought, discourses, and problematisations as it is about markets, relationships to the means of production, modes of accumulation, and class relations. These elements are incapable of becoming embedded through market forces alone. Rather, as Fernand Braudel (1981; 1982a; 1982b) argued, capitalism requires an active state to develop, protect, and provide nourishment for markets while mitigating the inherent contradictions that market relations facilitate. Similarly the ways in which the perceived imperatives of capitalism shape – and are shaped by – meanings that arise beyond the immediate economic realm are important. Culture – specifically the socio-political imaginaries of what is natural that culture fosters – is central to the perpetuation of any form of economic organisation, including capitalism. Third, such a model necessarily contends with problems of temporality. In a complex assemblage, not all elements run at the same pace, are positioned on the same temporal plane, or move along similar trajectories. Older elements intermingle with the new as well as those that are in a state of becoming. Elements may be absent only to re-emerge when least expected. Their presence may become possible by certain confluences of conditions and not others while at the same time fostering or annulling their influence. Thus, while analysts often wish to speak of distinct 'eras', these most often are a pastiche of old, more recent, and emerging mechanisms, forming a palimpsest upon which the various scripts of power-relations have, are, and will be written.

The claim here contrasts to the studies of nineteenth-century capitalist development noted above where military rationalities were argued to have been transposed into the civil–commercial realm. My argument is that in addition to scientific advances – which have been shown to be socially embedded – war-fighting doctrines are currently influenced by a contemporary capitalist regime of truth and style of thought. According to Foucault (1977: 131), every society has:

...its regime of truth, its "general politics of truth": that is the types of discourses which it accepts and makes function as true; the mechanisms and instances which enable one to distinguish true and false statements, the means by which each one is sanctioned; the techniques and procedures accorded value in the acquisition of truth; the status of those who are charged with saying what counts as true.

These form the passages through which arguments must be channelled in order to have a chance of being considered within policy discussions. Within the culture of contemporary capitalism, the regime of truth is constituted by methods of truth adjudication that derive from a particular understanding of classical economics that stresses the moral primacy of markets. Economists, business analysts, efficiency experts, and those who study the dynamics of organisational forms are those most often vested with the authority to speak truth. Markets are positioned as exogenous to social structures. Rational choice theories and new public management assumptions have become the beacons used to discern 'market realities' from what are presented as discredited forms of state economic management. A narrowly construed short-term understanding of efficiency is the metric used in order to evaluate the success and moral value of any public policy as opposed to justice, inclusion, or equality. Ontologically, markets are privileged over other systems of allocation and are invested with the qualities of a natural force. They are represented as possessing the ability to enforce their will and discipline on recalcitrant bodies – figuratively and literally – through the material effects of their logics. The precariousness created by the capriciousness of markets and attempts to mitigate uncertainty through deregulation and flexibility are asserted to be the very essence of freedom. Moreover, competitiveness and growth – both in absolute and relative terms – have become embedded into the policy imaginary. Whereas the state was once viewed as a provider of social protection against the externalities of markets when these externalities had been realised – what Karl Polanyi (1944) referred to as the 'double-movement' – this role has now shifted into serving as an enabler of market forces and facilitator of private profit-seeking activity.[5] Given the breadth of processes, policy, and power-relations underpinning the predominance of markets, a range of terms have been used to describe the current junctures of capitalism including the Washington Consensus (Williamson 1993), disciplinary neoliberalism (Gill 1998), market liberalisation (Stiglitz 2000), post-Fordism (Amin 2008), post-modernism (Harvey 1989), and globalisation (Beck and Camiller 2000).

While the renewed emphasis on markets has significantly affected public policy – as will be shown below with regards to defence procurement – and private commercial activity, there has been an accompanying cultural shift that Sennett (2006), Stephen Gill (1995), Naomi Klein (2007) and others argue has profound implications for governance and society. As Gill (1995: 399) has noted, the structures and language of social relations – and war is a social relation – are now conditioned by the 'commodity logic of capital', infusing market norms into the very practices of everyday life. Of primary concern for the analysis here are the following inter-related phenomena that contribute to this logic: speed, information, flexibility, efficiency and automation as well as the dynamics that flow from them: surveillance, deskilling, delayering, and spectacle. What will be demonstrated is how these values are constituted in contemporary capitalism. This is not to say that these values are necessarily adhered to or fully realised. Rather, the point is that they serve as touchstones within the contemporary regime of truth; they demand genuflection and consideration even if their implementation is ultimately partial, inconsistent, flawed, or contradictory. Subsequently, how these particular values resonate in RPA discourse and its associated war-fighting practices will be discussed in relation to the historical development of this technology.

The first important development conditioning the commodity logics shaping the rhythms of everyday life is the primacy of information and speed (e.g., Kitchin and Dodge 2011; Virilio 2005; Sutherland 2014). Although always privileged under capitalist market relations, information and speed, both as commodities and processes, have an increased importance in the organisation of society and commerce. The intrinsic and market value of information has rapidly increased as digital technologies facilitate the collection, analysis, and distribution of information in ways previously not thought possible (Harcourt 2014). An increased volume of information is almost always considered to be better and various inducements, tactics, and clandestine means of obtaining data have been recently developed from store loyalty cards to social networking websites. In addition, the speed with which information can be collected, analysed, and acted upon has been reduced to a measure of nanoseconds in some areas of activity (e.g., trading). Acting instantaneously upon information obtained in real-time is now the ideal decision-making model (Vostal 2014: 98). The commodification of information and speed as well as the processes through which both are generated have been greatly facilitated by the convergence of networks and communications technologies (e.g., Agger 2004; Hassan 2009). The sourcing and accumulation of information flows as well as the development of the rapid fire means of discriminating useful from non-useful, precise from imprecise, and actionable from un-actionable information is a defining problematic of contemporary capitalist cultures.

It is sometimes argued that the proliferation of information, the means by which to gather it, and the speed with which it can be analysed and shared is empowering to those who might otherwise be subject to coercion or control (for a discussion see Davies 2012; Davies and Bawa 2012). These dynamics are linked to the exercise of autonomy, decentralisation, democratisation, and the prevention of authoritarian tendencies. But rather than contributing to a decentralisation of forms of authority and management, Richard Sennett argues that the inverse has taken place. Whereas directives were once modulated through the interpretation of subordinates, the information revolution has established 'a new kind of centralisation' that relies on the capability to rapidly disperse commands directly, monitor their implementation, and adjust them according to the results achieved (Sennett 2006: 43). With the increase in the ability of managers to monitor subordinates in real time and to measure their relative rates of productivity has come an increased pressure to perform above expectations. Even professional occupations are unable to escape the purview of the new managerialism, its obsession with targets, and the forms of internal discipline that it fosters through systems of surveillance that are difficult to elude.

The increasing capacity to collect and assess information in real-time has also contributed to concurrent changes in the design, structure, and function of organisations and the practices of production. Whereas firms once organised themselves around principles of mass production, specialisation, economies of scale, and stable labour markets said to characterise Fordist models of production, the current model is one that emphasises flexibility in terms of production, tasks, labour, and organisational form. Ash Amin (2008: 2) via Sternberg has referred to these developments as 'flexible specialisation' in an economy defined by '...specialist units of production, decentralised management and versatile technologies and workforces, to satisfy increasingly volatile markets.' As Sennett (2006: 48) notes:

> ...the flexible organisation can select and perform only a few of its many possible functions at any given time...the sequence of production can be programmed in any sequence...[and] labour is task oriented rather than fixed form.

This shift has resulted in practices such as 'just-in time production' that seek to reduce the costs associated with inventories, materials, and over-production by tying manufacturing processes directly to market demand. Within a flexible system of production, sharing information across the firm, establishing efficient means of sourcing inputs, combining them into finished products, and distributing them in such a way that consumption is seamless is important. While the structural form of these systems remains a source of contention – i.e., are they linear chains, non-linear-circuits, or interdependent webs – their impacts have been shown to be both significant and complex (e.g., Hughes and Reimer 2004). The commodity chain has obtained significant metaphorical value in contemporary politico-economic discourses. It is not only a significant aspect for logistical configurations but also a unit of analysis for those seeking to challenge contemporary commercial practices (e.g., Barrientos 2013; Rossi et al. 2014).

Shifts in the organisation of production have also affected the position of labour. Fordist industrial mass production was in part said to be premised on employer concessions on job security, trade union membership, improved remuneration, and benefits that sought to capture a stable pool of workers in a fixed space. In contrast, it is argued that contemporary firms and service providers build their organisational structures around embedding precarious labour into the production process (Kalleberg 2011; Lee and Kofman 2012).

Temporary labour, term-limited contracts, few – if any – benefits, and a reduction in salary levels relative to senior management are now the new norms (Madrick 2012). It is claimed by industry that these changes better motivate workers, reduce employer overheads, reduce unemployment, improve efficiency, and hence profitability (Reilly 1998; Pyper and McGuinness 2014). Governments have shifted macro-level labour policy as a result. For example, parts of Europe – including the Nordic countries – have championed 'flexi-curity', a mixture of labour market deregulation and a social safety net designed to catapult individuals back into the labour market – often through negative forms of incentivisation such as reduced unemployment benefits for the able-bodied of working age (Wilthagen and Tros 2004). Impetus for these programmes – and concurrent pressures for workers to accept less – is fostered by the presumed ease with which firms can re-locate operations off-shore to benefit from more favourable (i.e., lower wage and even more highly deregulated) labour markets.

Within a system premised on flexibility, workers are less likely to be life-long specialists. This can be thought of in terms of what was once identified as having a trade or craft under pre-industrial forms of production but also encompassing the clear delineations of specialisation or role under Fordist modes of mass production (Kirpal 2011). Instead, it is argued that contemporary workers – whether in the office, shop floor, or sales floor – are required to multi-task and quickly adapt to changing demands and needs brought on by market forces – both direct and indirect (e.g., Doogan 2001; Thompson 2013). The impacts of these changes on labour – beyond the increasing precariousness of working conditions and shifts to 'multiple career careers' rather than lifetime employment in a specific industry – have been more contested. Some have argued that the industrial revolution led to deskilling as the mass production of goods removed the profitable space within which craft or small scale production could take place.[6] Workers went from being tradespeople with a deep set of skills in a particular area such as furniture-making to cogs within an industrial machine.[7] The deskilling argument today continues but in a slightly different form (e.g., Hassard et al. 2012; Fearfull and Dowling 2011). No longer is the claim made on the basis of a lack of breadth but rather focuses on a lack of depth. With demands made of workers – manual and professional – to undertake an increasing variety of roles for varying amounts of time with a changing volume of prioritised tasks on an as needed basis, it is argued that specific knowledge and depth of skills across these fields becomes more shallow (Spohrer et al. 2010). Thus, while the expectation may be that an employee shows a limited competency across a range of fields, their ability to rapidly adapt may be seen as more valuable to a firm than demonstrating deep competence in any one field in particular (Beckett and Hager 2013).

Concurrent to these changes in how labour is mobilised within the firm have been practices that seek to minimise the total number of staff directly employed – whether on permanent or temporary contracts. Leanness has become a value in and of itself. The minimisation of labour has been achieved through two methods. The first is the outsourcing of labour to independent firms who offer to provide services for a total cost that is lower than directly employing workers (Levy 2005). The reduction in costs achieved by firms that provide outsourced labour may be a function of non-unionised environments, flexible contract arrangements, geographic location, deregulated labour markets, lack of benefits, poorer health and safety regulations, currency differentials, or combinations of any of the above (Sengenberger 2005). Outsourced labour is extremely flexible in that the number of workers required can be quickly tailored to production demands without any overhead costs being borne by the contracting firm. Moreover, the use of outsourced labour provides the potential for firms to take advantage of structural inequalities – socio-economic, geopolitical, or even regional – without being held directly accountable for exploitative labour practices. The second method has been called delayering. It refers to a process of organisational restructuring that seeks to eliminate as many layers of 'middle management' as possible (Collinson and Collinson 1997). Gaining prominence in the 1990s – in part because of the developments in information capture and speed noted above – delayering was not only argued to make firms leaner and more efficient by eliminating redundant levels of managerial oversight, but the practice was also said to contribute to normative goods such as autonomy and democracy by linking workers and site managers more directly to senior management (Beirne 2013: x). Such an organisational structure was said to increase the speed with which information gleaned from workers at the 'coal-face' could be analysed and acted upon by management.

To further efficiency, productivity, cost-cutting, and ultimately profitability, firms also looked to automate as many labour processes as possible. Unlike humans, automated processes were perceived as predictable, controllable, and certain in their parameters. Machines and computers are often faster, more powerful, and less expensive than their equivalent in human form (Levy and Murnane 2004). Moreover, as technologies became ever more sophisticated, the speed with which certain key tasks could be accomplished surpassed what could be achieved through human labour alone. Thus, while forms of simple automation had taken place for several decades in manufacturing, the rapid improvements in computer technology that began to cascade in the 1970s significantly transformed capitalist cultures. On the one hand, information technology potentially called for a rethink of core tenets of what capitalism might entail with respect to ownership of the means of production, as a variety of occupations began to make increasing use of computer technology that could be owned by individuals. On the other hand, the development and widespread adoption of computing technology and processes of automation led to two important structural changes. The first was the creation of what Brian Arthur has called the 'Second Economy', a space where computers communicate with other computers in order to conduct transactions across networks, particularly in service industries that were once heavily dependent upon human labour, including diverse sectors like online shopping and publishing (Davidow 2012). While highly productive and currently enjoying rapid expansion, Arthur argues that an important consequence of the Second Economy is that it requires fewer workers, particularly 'lower-productivity' service workers than traditional sectors of the economy. The second economy thus feeds into the second structural change or what Sennett (2006: 92) has referred to as the 'spectre of automated uselessness'. Not only are workers – and employers – aware of the potential for any form of labour to become automated but that automation itself is decoupling job creation from economic growth.

A system of production where information, speed, flexibility, efficiency, and automation are accorded cultural value is still reliant on consumption for its reproduction and expansion. However, it is argued that the role of consumption has transformed within the contemporary capitalist cultures into 'an immense accumulation of spectacles' where social relations amongst people are 'mediated by images' (Debord 1995). These relations have transformed from being primarily oriented around *being*, to *having*, and now *appearing*. Guy Debord (1995: 29–31) argued that the spectacle is thus '...the ruling order's nonstop discourse about itself, its never-ending monologue of self-praise, its self-portrait...[that] depicts what society *could* deliver, but in so doing it rigidly separates what is *possible* from what is *permitted*'. The values that have led to the changes in production and the power of labour noted above have combined with the impetus to manufacture the *appearance* of *having*: success, wealth, leisure, happiness, security, and power. Debord (1995) argued that the end result is increasing isolation and alienation as people are unable to divorce themselves from 'dominant images of need', the identification of new desires, and a definition of satisfaction that constantly expands outwards to encompass new commodities – not so much for what they do but for what they are said to represent. Sennett (2006: 154) argues that a key part of the dynamic is that the capabilities of commodities have been emphasised at the expense of how they will actually be used. The end result is that consumption exceeds limits that might be imposed by one's own capabilities to utilise what is being consumed: 'you don't limit what you want to what you can use.' But as will be shown below, these changes have travelled beyond relations of production and consumption.

Science

Many of the values of contemporary capitalist cultures resonate within leading fields of scientific discovery – and vice versa. As discussed above, this observation is not a claim for a casual account of how ideas, problematics, theories, or discourses circulate within one domain identified as a prime source and then migrate to others as derivatives. Rather, the argument is that points where economic organisational activity and science converge as values in one discourse may also resonate in other discourses. In particular, the primacy given to flexibility, adaptation, information, and speed as values in contemporary capitalist cultures have also been identified by Bousquet (2009) as focal points of scientific inquiry into chaoplexity and networks. These areas of study have in turn influenced theories of contemporary warfare (Guha 2011: 110–132).

Motivated by the challenge of attempting to discern meaningful patterns in non-linear systems which exhibit positive feedback characteristics, studies of chaos and complexity (chaoplexity) emerged in the second half of the twentieth century.[8] Through the observation of various phenomena and the development of non-linear modelling, it was clear that linear systems of organisation in the natural world were the exception rather than the rule. But the contribution of chaos theory to science was the claim that one could still '...identify a structure and order to phenomena which previously appeared to have none', though these were often extremely complex (Bousquet 2009: 169). Non-linear modelling and simulation made possible by advances in computing technology demonstrated how the hidden order to disorder had a '...*sensitive dependence on initial conditions*' (Bousquet 2009: 171; italics in original). To better understand complexity, the concept of the network was linked to the qualities emphasised by chaos theory (i.e., non-linearity and positive feedback) in order to understand patterns that might emerge through processes of self-organisation. In addition, complex adaptive systems – systems that transform by learning from experience – were identified by key properties including decentralisation, dispersal, and internal competition and cooperation which in sum produced the overall behaviour of the system (Bousquet 2009: 175). Complexity science thus began to hypothesise and demonstrate that '...a small change made to the parameter or control values of a system could cause a sudden qualitative change in the system's long run dynamical behaviour' (Bousquet 2009: 176). The important strategic point was that just prior to system bifurcation (i.e., the point where a system could transform into a new but unpredictable order), that is the stage where the potential for bifurcation was greatest but stability could still be maintained, was the optimal point for system flexibility and adaptability. At the 'edge of chaos', such systems were argued to be best suited to responding to contingency and unpredictability (Bousquet 2009: 183).

Several key generic prescriptions for organisational systems thus came forward from research into chaoplexity. The first was that networks, '...decentralised, open, and adaptable forms of organisation', were to be the preferred structure in order to harness efficiency and dynamism (Bousquet 2009: 205). Within these networks, it was argued that elements:

> ...should be loosely connected together with a built-in redundancy and ability to reconfigure their positions within the network when necessary, allowing for the emergence of new behaviour and organisational arrangements.
>
> (Bousquet 2009: 202)

Therefore, arrangements whose organisational form straddled the edge between order and disorder were argued to be '...naturally best suited to adjusting to a rapidly changing environment through the self-organising and emergent properties of the network' (Bousquet 2009: 205).

Bousquet is able to demonstrate the specific impact that chaoplexity and network thinking have had on contemporary war-fighting doctrines. While analysts often focus on the network characteristics of terrorist groups (e.g., Fellman and Wright 2004; Ilachinski 2012), large militaries with technologically sophisticated weapons systems have also begun to search for ways to harness the capabilities of networks (e.g., Moffatt 2010; Mitchell 2013; Pang et al. 2012). The objective is to be able to readily transform as conditions change without causing systemic disruption. This flexibility is to be achieved by facilitating the ability to exchange information (and the speed with which this can be accomplished) between parts – or what is sometimes referred to as 'jointness' or 'synergy' (Bousquet 2009: 205; Sloan 2012: 50–61). In this regard, specific metaphorical and conceptual significance has been attached to the notion of the 'swarm', '...the networks of distributed intelligence that enable bees, ants, and termites to evolve complex forms of behaviour on the basis of the simple rules of interaction of their individual members' (Bousquet 2009: 210). It is argued that swarms are resilient, flexible, and quick to adapt to changing conditions (Cevik et al. 2013; Henkin 2014). This empowers swarms to adopt innovative and novel tactics in response to changes in the operational environment (Meiter 2006: 203). Moreover, it is claimed that given the networked form of contemporary security threats from terrorist insurgencies to cyber-warfare, swarming is more suited to the shape of contemporary warfare than traditional command and control models (Arquilla and Ronfeldt 2000). But as Bousquet notes, this does not signal the end of hierarchy within military organisations. Rather, much like in the contemporary firm where flexibility is desirable but management is still considered essential, the goal is to develop hybrid organisational forms that '...harness the flexibility and adaptability of networks while preserving some hierarchical features' that might foster 'control and predictability' (Bousquet 2009: 210).

In the next section, how the discourses of chaoplexity and networks have operated alongside congruent values found in contemporary capitalism to shape modern war-fighting will be examined. The intersections of science, capitalism, and war will be mapped in the recent changes to US military doctrine. The US has been selected for illustrative purposes as it is both a predominant military power and market oriented in its economic organisation. The argument presented is that the way in which contemporary war-fighting is being problematised has been shaped by the structural reorganisation of the US military through military transformation, the Rumsfeld doctrine, and the understanding that adopting networked forms are the most effective means of combatting networks that may pose a threat. Thus, the new prominence of the RPA in military thinking and procurement is not simply an effect of technological advancement and availability. Rather, it has been made possible by a complex (and at times potentially contradictory) regime of truth that fosters information gathering, flexible structural forms, automation, speed, and spectacular consumption.

Revolution, military transformation, the Rumsfeld doctrine, and the dawn of netwar

Conventionally, the revolution in military affairs (RMA) has referred to the drive towards developing and deploying advanced weapons and information technology, improving command and control systems, engaging new forms of war fighting that focused on producing effects which rendered notions of discernible lines obsolete (Cohen 2004: 395).[9] Advocates like William Perry argued that the RMA would increase the combat effectiveness of any single technology by fostering synergies amongst platforms through:

> ...communications, computers, command, control, and intelligence processing (C4I); stealth or low-observability; precision guidance; and intelligence gathering, surveillance, and reconnaissance (ISR).
>
> (quoted in Sloan 2012: 50–1)

Others like William Owens and John Shalikashvili emphasised the promotion of inter-operability, communication, and the 'seamless integration of service capabilities' across military systems and platforms (Sloan 2012: 54–55). Both effectiveness and efficiency considerations in strategic and financial terms – with regards to reducing the number of redundant systems across the armed forces in an era of perceived fiscal constraint – provided justifying logics. Reaching its zenith in American military circles in the early post-Cold War era, RMA systems were argued to contribute the capabilities necessary to fight two large-scale theatre wars as well as maintaining a significant technological advantage – 20–30 years in real terms – that would provide a credible deterrent against any emerging threat that might harbour hostile intent. Simply put, it was argued that the deployment of high tech systems would be able to dismantle an inferiorly equipped force with ease, an argument that was bolstered with the impressive results of the first Iraq War and the ways in which C4I, stealth, precision, and ISR technologies were represented as essential to the comprehensive defeat of Iraq's voluminous forces.

This hubris over the continuation of American military dominance because of technological superiority was not free of dissent. For example, in what would become one of the most highly influential works of military strategy of the late twentieth century, John Arquilla and David Ronfeldt (1993: 25) warned that, 'technology permeates war but does not govern it. It is not technology per se, but rather the organization of technology, broadly defined, that is important.' With the information revolution they argued that war was in the process of being reformulated in such a way that superior military forces would be those that:

> ...are well prepared, make room for manoeuvre, concentrate their firepower rapidly in unexpected places, and have superior command, control, and information systems that are decentralized to allow tactical initiatives, yet

provide the central commanders with unparalleled intelligence and "topsight" for strategic purposes.

(Arquilla and Ronfeldt 1993: 2)

From this perspective, the first Gulf War was to be seen as the ending, rather than the beginning, of an era in terms of how threats were organised and mobilised. It was argued that netwars, wars in which information and communications were prime areas of contestation, were in the ascendancy. The implications would be considerable for the traditional economic, political, and military dimensions of war-fighting. And implicit in the argument being forwarded by Arquilla and Ronfeldt (1993) was that non-state actors such as terrorists, organised crime, and drug syndicates were more capable of engaging in the warfare enabled by the information revolution than the closed and rigid hierarchies of traditional military structures.

At the time, these views were considered highly unorthodox; however, the events of 9/11 challenged the orthodox world-view of nation-state warfare being assumed within the parameters of the RMA. Conventional war-fighting against an identifiable and conventionally configured opponent whose intentions would be known could no longer to be taken as the norm in grand strategy. Within this emerging context of ambiguity, US Secretary of Defence Donald Rumsfeld (2002: 23) argued that:

> [the] challenge in this new century is a difficult one: to defend our nation against the unknown, the uncertain, the unseen, and the unexpected...to accomplish it, we must put aside comfortable ways of thinking and planning – take risks and try new things – so we can deter and defeat adversaries that have not yet emerged to challenge us.

Within defence discussions, transformation – as opposed to the RMA – became the nodal point through which strategy, tactics, and organisational structure were developed, debated, and deployed. Rumsfeld believed that transformation would be central to maintaining the predominant position of the United States within the emerging security environment. Thus, he oversaw a shift in US military grand strategy away from a two-theatre war construct based on threat assessment to a 'capabilities-based' approach that emphasised how threats might emerge and what would be needed to deter and defend against them. Concurrently, rather than maintaining a force size and structure that would be capable of two simultaneous occupations, Rumsfeld (2002: 24) moved to a doctrine that sought the capability to swiftly defeat two aggressors at the same time while maintaining the option of a 'massive counteroffensive to occupy an aggressor's capital and replace its regime'. This shift in outlook was proposed as a means of augmenting the US capability to deal with contingencies that would arise through the adoption of the 'capabilities-based' approach to risk assessment. As Maria Ryan (2014: 46) has noted, this preference '...did not mean that the Pentagon was abandoning its commitment to conventional military superiority'. Rather, as the war on terror developed and it became increasingly untenable to frame the insurgency in Iraq as a conventional war-fighting force, Rumsfeld realised that it was necessary 'to complement conventional strengths with IW [irregular war] capabilities so that...[the US]... could fight and win across the full spectrum of conflict' (Ryan 2014: 46). Full-spectrum dominance across all types of warfare was to be the goal.

But underlying this emphasis on the ability to adapt in 'a world defined by surprise and uncertainty' (i.e., a world shaped by chaoplexity) was Rumsfeld's own conviction that the institutional structure and organisation of the American military also required transformation (Rumsfeld 2002: 22). He asserted:

> We must transform not only our armed forces but also the Defense Department that serves them – by encouraging a culture of creativity and intelligent risk taking. We must *promote a more entrepreneurial approach*: one that encourages peoples to be proactive, not reactive, and to behave less like bureaucrats and more like venture capitalists.
>
> (Rumsfeld 2002: 29; italics added)

Thus, Rumsfeld accelerated processes begun during the 1990s to make the US military leaner through delayering and down-sizing, practices that were shown above to be embedded into core values that shape the cultural ethos of contemporary capitalism. While the use of private security contractors in Afghanistan and Iraq grabbed media and scholarly attention, Chalmers Johnson (2003: 53–58) has documented how from the 1990s, 'the Pentagon contracted out every conceivable kind of service except firing a rifle or dropping a bomb'. These support contracts grew to include construction, maintenance, security, cleaning, logistics, training, policing, intelligence, and food preparation in American bases and outposts around the world (Deitelhoff and Geis 2009: 8). For Rumsfeld, the imperative to embody these values was a central component to successfully implementing a capabilities-based approach to war-fighting and transforming the US military. He echoed Arquilla and Ronsfeldt when he stated that 'all the high tech weapons in the world won't transform the U.S. armed forces unless we also transform the way we think, train, exercise, and fight' (Rumsfeld 2002: 29).

While grappling with chaoplexity on the battlefield, the rise of netwar, and the privatisation of front-line support operations in the war on terror, deep structural changes were taking shape in the US military. As Armin Krishnan (2008: 3–4) argues, while the majority of attention focused on private security companies engaged in combat roles, this was actually marginal in comparison to the 'outsourcing of military technical services'. For example, beyond the sheer number of weapons systems that relied on private contractors for support, Deitelhoff and Geis (2009) argue that the doctrine underlying outsourcing itself also transformed. While it was once advocated that forces should develop organic capabilities that would allow them to operate and maintain any new weapons system that was deployed into service, new thinking argued that maintenance should be outsourced from first principles: after four years for critical weapons systems and for the life of non-critical systems (Deitelhoff and Geis 2009: 8). As a result of the modification in institutional mind-set regarding outsourcing, a mind-set that was intensified under Rumsfeld, the ratio of US military personnel to private contractors moved from 50: 1 in 1991 to 10:1 in 2003 to 6:1 in 2008 (older figures, Isenberg 2007: 83 quoted in Deitelhoff and Geis 2009: 9).

As with the more general changes in the structures of firms and labour noted above, the justificatory logic for the reorganisation of the American military mimicked cultural values embodied in contemporary capitalism noted above. The contracting out of military functions was argued to increase flexibility and improve the speed with which one could mobilise forces (Schwartz and Church 2013). It was claimed that in addition to these achievements, restructuring would also reduce the overall costs of maintaining a military force and deploying it for combat operations (Stanger 2009; Deitelhoff and Geis 2009: 9). But these arguments went beyond the strictly economic to encompass the strategic and the political. One supporting line of argument claimed that contracting out would shrink the commitment-capability gap that emerged with troop reductions at the end of the Cold War and the increasing number of missions/interventions requiring American troop deployments (see Baum and McGahan 2013). Another argued that outsourcing was politically expedient insofar as it potentially removed direct responsibility for the consequences of military actions away from government and onto private contractors (Godfrey et al. 2014: 113).

These lines of argument – particularly that outsourcing improves flexibility and ensures that capabilities match commitments – are premised on another core assumption constitutive of the culture of contemporary capitalism: that the private sector is more efficient than the public sector by definition. But the immediate economic case based on bottom lines may be secondary. Deitelhoff and Geis (2009: 11) argue that while the discourse of market logics and neo-liberal economics have been used to 'sell' outsourcing to politicians and the general public, it is the utility of private contractors '...as a flexible policy tool in pursuing security policy' that has given outsourcing lasting traction. Thus rather than seeing outsourcing as a weakening of the state's monopoly on the use of force, privatisation and outsourcing represent a '...variation in the state's exercise of it' (Deitelhoff and Geis 2009: 12). At a time when key policy-makers and military strategists began to stress the uncertainty, adaptability, and destructive potential of networked threats, the attraction of adopting an equally flexible organisational structure to neutralise these threats was strong. It is in this policy context – a context shaped by scientific thinking, the dominance of specific economic values, and changes to war-fighting doctrines – that the RPA emerged as a central tool in global counter-insurgency operations. In the next section, how these elements have come together to make possible the privileging of the RPA will be presented.

The RPA: conditions of possibility

The RPA is not a new technology. Early systems and associated doctrines were initially developed during World War I and II (Irvin 2003; Williams 2011; Hall and Coyne 2014; Gregory forthcoming), though precursors can be traced back to the 1780s in Europe with surveillance undertaken from hot air balloons by the French and over 2000 years ago in China for communications via floating air-lanterns (Clarke 2014: 231). More advanced systems were designed for fighter target practice in the 1950s. Advances in research and development meant that by the 1960s, the RPA began to be used for intelligence gathering missions over China and Vietnam (Webb et al. 2010: 31). In addition, the Ryan Firebee Unmanned Aerial Vehicle (UAV) was deployed as a low cost means of flooding enemy radar systems and air defences. From a humble beginning as fodder for target practice, RPA development, production, and utilisation proliferated as firms sought to meet the demands of the military (Hall and Coyne 2014: 450–457). Large platforms like the RQ-4/MQ-4 Global Hawk were developed for command, control, communications, computers, intelligence, surveillance, and reconnaissance (C4ISR) tasks over large territories while smaller systems like the RQ-14 Dragonfly – at 2.2 kgs – have been deployed by ground troops to scout locations in the field. But despite a wide range of systems being developed and procured, it is two RPA platforms in particular that have grabbed public attention in contemporary counter-insurgency operations: the MQ-1 Predator and MQ-9 Reaper.

The MQ-1 Predator was designed and developed by General Atomics based in San Diego, California. Over its 15 years of service in the US military, the Predator has evolved from a tool used in surveillance and intelligence gathering to a unit capable of destroying ground targets when required. Its original intelligence-surveillance–reconnaissance mandate has remained its primary *raison d'etre*; however, the weaponisation of the MQ-1 achieved with the addition of two Hellfire missiles, led it to being known as the 'killer scout' (Tirpak 2007: 47). This combat-capable version was first used by the Central Intelligence Agency for missions over Afghanistan in October of 2001. By 2007, the MQ-1 had accumulated in excess of 100,000 hours of flight over Iraq and Afghanistan and by 2011 it had surpassed over 1 million (Edwards 2007: 34; Balle 2015). Extremely lightweight at approximately 1000 lbs, the MQ-1 enjoys a range of 675 nautical miles at a cruising speed of 160 km/h at 3000m above ground (IISS 2009: 1–2; Edwards 2007: 34). While the small size and light weight of the MQ-1 provide advantages on surveillance missions, its slow speed and low ceiling make it vulnerable to very basic anti-aircraft systems. Thus the success and notoriety of the MQ-1 is part a product of a battlefield environment within which insurgents do not have widespread access to anti-aircraft defence technology. RPA losses to enemy fire have thus been relatively sparse. For example, from 2002 to 2005, the USAF lost an average of six drones a year (Edwards 2007: 34). But while the flight capabilities of the MQ-1 may seem distinctly ordinary in comparison to other aircraft available today, the sensor technologies on offer and the ability for this RPA to be controlled in real-time by pilots and operators thousands of miles away are the unique selling points.

The MQ-9 was developed as a larger and more robust off-shoot of the Predator capable of withstanding intensive combat missions. Known as the 'hunter-killer', the MQ-9 is a 6000 lb airframe capable of carrying a payload equal to that of an F-16 fighter jet (Tirpak 2007: 47). Typical configurations include nine Hellfire missiles, two 500 lb joint direct attack munitions, and two Sidewinder missiles. The MQ-9 is also capable of carrying laser-guided bombs or other munitions. For surveillance and targeting purposes, the Reaper carries sensor technology comparable to an F-16's sniper or litening targeting pods. In addition, the MQ-9 is fitted with Lynx Synthetic Aperture radar that provides all-weather, day-night, capability. The radar system works in conjunction with an additional sensor ball that contains daylight TV, low light intensified TV, and infra-red cameras that can track targets across a range of conditions. It is claimed that this technology provides operators of the Reaper with the video capability to spot an aerial rotating on a vehicle from 7 miles away (IISS 2009).

The MQ-9 has a range of 1000 nautical miles and normally operates up to 25,000 feet, making it possible to avoid anti-aircraft artillery and portable anti-aircraft missiles (IISS 2009). The Achilles heel of the Reaper is its relatively slow flight speed (approximately 165 mph) or 450 km/h. However, this is still faster than the Predator which can only reach one-third of this velocity. With its larger fuel carrying capacity, the MQ-9 is capable of remaining in station for up to 24 hours minus transit time. The ability to conduct extended surveillance has led one analyst to remark that with the Reaper 'the attraction is its persistence' (Tirpak 2007: 47). All of this persistence is argued to come at a relatively low cost. While the final price tag on a MQ-9 has been claimed to be about $7 million US, about twice as much as a Predator, this represents a 75 per cent savings on the purchase price of an F-16 fighter jet (approximately $30 million US). With this platform, the standard combat air patrol consists of four Reapers outfitted with sensors, a ground control station, communications equipment, and 30 support personnel per airframe. It has been claimed that the total cost of such a system is approximately $53 million (US), representing a savings on the operational costs of an equivalent piloted aircraft (IISS 2009). However, since being introduced into public discussions, these figures have been disputed. For example, Wheeler (2012) has argued that these estimates do not include development and other costs which, if added, would price a Reaper quartet closer to $120 million (US). Similarly, data provided by Boyle (2012) shows that the total labour required to keep systems like the MQ-9 operational may have been underestimated by nearly two-thirds.

Beyond perceptions of lower procurement and operating costs – that may not be matched in practice – the advantages of RPA technology are said to be numerous and wide-ranging. In the *2009 Roadmap for UAS Integration*

(RM2009), a defining policy statement on the aims, objectives, and future planning of RPA technology in the American military, the United States Department of Defence (2009: 2) argued that the RPA platform provides:

> ...a persistent and highly capable intelligence, surveillance, and reconnaissance (ISR) platform to troops requiring a look 'beyond the next hill' in the field or 'around the next block' in congested urban environments and, if necessary...[it can] also assist troops in contact or perform strike missions against high value targets (HVTs) of opportunity.

It is argued in the report that the RPA offers 'versatility persistence' (DoD 2009: xxiii). In particular, the RPA is able to extend and deepen the line of sight for troops and their commanders, both at the operational and tactical levels, through the provision of enhanced levels of vision – and different vantage points across visual spectrums. The provision of a viewpoint above the fray is considered to be a significantly useful attribute. The flexibility of the RPA is also highlighted in the *RM2009*. Its operational tasks can be re-assigned dynamically while in a battle-space and it possesses the capability to 'operate beyond line of sight (BLOS)'. These attributes have made it possible for pilots and sensor operators to fly missions from well outside the zone of immediate danger while 'maintaining only a small contingent forward in the operational environment' (DoD 2009: 3).

This flexibility translates into how RPA platforms are deployed and the roles that they are asked to perform. According to an assessment in the *RM2009* that examined the suitability of RPA platforms to generic mission types, the technology had the potential to contribute to battle-space awareness (30 mission types), force application (22 mission types), protection (11 mission types), command and control (8 mission types), net-centric (8 mission types), building partnerships (6 mission types), logistics (6 mission types), and force support (2 mission types). These advantages have thus led to various forms of RPA being deployed by all of the main branches of the US military (DoD 2009: 8).

While it is estimated that at least an additional 50 nation-states are now deploying drones for various combat, reconnaissance, and surveillance missions, as an early pioneer in the field of RPA development, the American deployment remains the largest (Hastings-Dunn 2013: 1241). Estimates of the total fleet size in 2009 detailed 138 airframes split between 118 General Atomics MQ-1 Predators and 20 General Atomics MQ-9 Reapers with the air-based systems alone integral to 18 different mission types (IISS 2009). The *RM2009* reported at least 40 air-based RPA systems in service, in production, or in research and development stages. By 2012, a Congressional Research Services Report on the unmanned (sic) air systems (UAS) capability of the United States reported that the armed forces had nearly 7,500 platforms in service (from micro drones to large units) including 54 Reapers, 161 Predators, and 26 MQ-1C Grey Eagles (an upgraded Predator platform used by the Army) (Gertler

2012: 9). The centrality of the RPA to contemporary military activity and the number being procured and used in operations has increased since that time. The *RM2013* recorded nearly 11,000 RPAs in service with an inventory of 237 Predator-Grey Eagles, 112 Reapers, and 28 MQ-8 Fire Scouts – a remotely piloted helicopter platform capable of applying kinetic force (DoD 2013: 5).

RPA acquisition

As irregular warfare has become increasingly prominent in the operations being undertaken by Western militaries as a part of the global pacification efforts, the RPA has been identified as an important element of counter-insurgency war-fighting. This reflects a deeper desire to replace human operators and soldiers with machines capable of decision making in complex environments. As in civilian forms of production, the belief is that these systems will be more effective, productive, efficient, and precise (Singer 2009). Moreover, within the combat context, the RPA is perceived as reducing the potential to suffer the kinds of casualties that are politically costly and militarily demoralising. Nowhere is this more apparent than in the US military. Previously, Edward Helmore (2009) reported that plans to 'robotise' around 15 per cent of the military were already underway. In 2009, there were over 5,000 robotic vehicles and drones deployed in Afghanistan and Iraq. With some hyperbole, it has been argued that the momentum for the procurement and deployment of drones means that 'the end of the era of the fighter pilot is in sight' (Helmore 2009).

The impetus to develop, commission, procure, and deploy the RPA in contemporary and future battle-spaces extends beyond perceptions of its operational and tactical utility. It has been institutionalised within the American legislative system as an enshrined priority with regards to defence spending and procurement. First, in Section 220 of the *Floyd D. Spence National Defense Authorization Act* for the 2001 fiscal year (Public Law 106–398), Congress made a specific demand on the DoD with regards to RPA development: by 2010 one-third of the aircraft in the 'operational deep strike force should be unmanned'. Over $100,000,000 in funding was dedicated to the task of acquisition assessment alone. This was later supplemented by the *John Warner National Defense Authorization Act* for the 2007 fiscal year (Public Law 109–364) which offered guidance on the future development of RPA and other unmanned systems technology. The *John Warner Act* enshrined a preference for unmanned systems in acquisitions of military equipment, demanded the joint development and procurement of unmanned systems and their components across services, encouraged moving service specific unmanned systems to joint systems as appropriate, promoted the development of more effective management structures, encouraged the coordination of the development and procurement of unmanned systems, and set into motion an auditing plan to assesses the progress towards meeting targets articulated in Section 220 of the *Floyd D. Spence National Defense Authorization Act* (DoD 2009: 4–5). Thus

RPA platforms were embedded into the wider goals of military transformation as well as the continuing influence of jointness and synergy from the RMA.

Legislation has thus institutionalised demand for RPA platforms in the American armed forces. But the United States is not alone and a lucrative market for aero-space contractors in the defence industry has been created. In 2007, global sales were estimated at more than $4.7 billion with 60 per cent of the market located within the United States. In 2014, it was estimated that only 11 states possessed combat-capable drone technology: the United States, France, Germany, Italy, Turkey, the UK, Russia, China, India, Iran and Israel; however, by 2022, it was expected that the global market would be worth $82 billion. The market share held by the United States was expected to shrink as overseas commitments decreased – with resulting budget cuts – while procurement by Russia, China, India, and Japan was set to expand (Medina 2014). Within Europe, leaders in research, development, and production have been the UK, France, Germany, Spain, and Italy. But a considerable amount of coordination has been undertaken by EU, which by early 2014, had dedicated €315 million for research and development, with a particular emphasis on platforms that could contribute to an integrated border security system (Hayes et al. 2014: 26). The UK in particular has also been dedicating an increasing amount of effort and resource towards improving its own RPA development capability. Of note is the Strategic Unmanned Air Vehicle (Experiment) Programme (SUAV(E)) and BAE Systems development of the MANTIS UAV, which has catalysed a search for partnerships with the Engineering and Physical Sciences Research Council, and the Taranis platform (Webb et al. 2010: 33). In terms of combat RPAs, there has been bilateral defence cooperation between the UK and France – though projects have so far fallen through – as well as between Spain and Germany with the Barracuda project. The development of the nEUROn RPA is a multi-lateral endeavour involving French, Swedish, Greek, Swiss, Spanish, and Italian firms (Hayes et al. 2014: 41–42). Outside of Europe, China also has been developing its drone capability with concerns that it is set to become a key proliferator of the technology (Hsu et al. 2013) while Iran has incentivised RPA development at the national level and demonstrated a drone with a claimed range of 2000 kilometres (Kreps and Zenko 2014: 72).

RPA deployment

Given the myriad issues that targeted killing by 'hunter-killer drones' raises and how their operations tap into the geopolitical imagination of the general public, it is little wonder that these platforms and their operations capture media attention. But it is worth noting from the outset that the American military and NATO stress the valuable role of RPA platforms and other unmanned systems in undertaking 'menial' jobs as opposed to their role in targeted killing. As one NATO special report stated:

> Unmanned [sic] vehicles are ideal for carrying out dull, dirty, and dangerous jobs: Robots do not mind circling the skies of Afghanistan for dozens of hours; contaminated environments, such as Fukushima, do not pose an obstacle; and losing a downed drone is a far smaller loss than losing a pilot with his [sic] aircraft.
>
> (Nolin 2012: 2)

Even for those RPA platforms tasked with operations within a battle-space, it has been claimed that strikes are relatively infrequent. For example, it was estimated by the IISS (2009) that the USAF made approximately 365 strikes using RPA platforms from 2007 to 2009. While this represents a large number in aggregate, in comparison to the overall volume of missions, it translated into weapons being fired in about 2.5 per cent of the sorties undertaken (IISS 2009). Meanwhile, it is thought that overt interventions in Afghanistan, Iraq, and Libya have led to over 1,200 RPA strikes by American and British forces (Woods and Ross 2012). Official numbers have been clouded in mystery, particularly since the decision undertaken by the US Air Force in 2013 to stop recording drone activity in its monthly air power summaries for Afghanistan, though leaked numbers showed RPA strikes accounting for one-fourth of kinetic force operations in January 2013 (Ross 2013).

Still, even though remotely piloted systems are being developed for a plethora of 'dirty jobs' from medical evacuation to IUD clearance, their potential use in combat continues to dominate contemporary discussions and their spectacularly imagined futures. In the *RM2009*, the Department of Defence (2009: 10) asserted the primacy of the war-fighting function of RPA platforms by stating that:

> today, Predator, Reaper and Extended Range/Multipurpose (ER/MP) UAS are weaponized to conduct offensive operations, irregular warfare, and high value target/high value individual prosecution and this trend will likely continue in all domains. In the air, projected mission areas for UAS include air-to-air combat and suppression and defeat of enemy air defense.

But the spectacularly imagined futures of RPA platforms and their associated desires go beyond the extension of combat capability. As mentioned above in relation to cultural values indicative of contemporary capitalism, in the world of manufacturing, streamlining commodity chains are emphasised as a means of reducing costs and increasing the overall efficiency of production. Thus, a key for business success has been to develop the systems necessary to quickly source raw materials and parts, manufacture and assemble products as required, and then distribute them on demand. The successful completion of these tasks requires an immense amount of coordination, information sharing, and speed across what is an increasingly global marketplace. Firms have adopted various means of achieving these aims, from highly centralised systems of demand evaluation and supply coordination to decentralised networks of flexible production and distribution based on local needs. In this way, the management of commodity chains reflects tensions outlined by Bousquet (2009) in contemporary military thinking regarding the desire for centralised command and control systems that use the speed of their processing ability to dictate the tempo of combat to a point where the adversary is incapable of keeping up, to those of decentralised chaoplexic war-fighting where decentralisation confers a degree of flexibility and innovation that confounds an opponent.

RPA platforms exhibit a similar tension in terms of the ways in which they are being deployed and the advantages they confer that go beyond identified technological capabilities. The Department of Defense (2009: xiii) has argued that:

> the fielding of increasingly sophisticated reconnaissance, targeting, and weapons delivery technology has not only allowed unmanned systems to participate in shortening the 'sensor to shooter' kill chain, but it has also allowed them to complete the chain by delivering precision weapons on target.

Similarly, the RPA is seen by the USAF as an important component in irregular warfare for developing the speed and range necessary to *compress* the find-fix-track-target-engage-assess (F2T2EA) elements of the 'kill chain'. This is particularly important to the kind of operations in which these systems are deployed. According to the Irregular Warfare Doctrine of the USAF (2007: 16), compression is vital when dealing with time sensitive and/or high value targets, the kind of missions often undertaken by the RPA. Thus, the attraction of the RPA becomes apparent given the primacy of tempo to the 'kill chain'. Their multi-functional capabilities allow several roles to be performed by one system and one set of operators. This is argued to provide the potential to eliminate time lags in communication, acclimatisation, and situational awareness that are inherent to switching platforms mid-mission. Moreover, the compression of the F2T2EA reduces the opportunities for targets to take countermeasures or evasive actions that may become possible if they become aware that they have been targeted. In the RPA 'kill chain', speed literally kills. Being able to seize upon dispatching opportunities in real-time as they arise thus is believed to confer significant advantages.

But as in most organisational systems that seek increased 'command and control', there are desires, including those articulated in procurement legislation, for improved coordination and integration amongst RPA platforms themselves. Initially, these desires were evident in attempts to implement a coordinated pack hunting system of hunter-killer teams. As documented by J.R. Wilson (2002), the aim was to have Predators provide reconnaissance, identification and laser targeting capability with Reapers directly engaging with targets. As a precursor to synergy through the coordination of drones, RPA platforms had been used in combination with manned aircraft, including the provision of real-time imagery to monitors aboard AC 130-U gunships. More recently, the goal has been to develop autonomous systems that would be fully integrated with one another. Forming a 'swarm', these drones would be capable of analysing the battle-space, performing complex tasks, engaging in decision making about how to best accomplish tasks, and adapting to and learning from conditions and outcomes. Human operators would assume the role of a coordinator, ensuring that packs of drones – each with specific mission directives or roles – were operating at maximum efficiency. Already the Global Hawk RPA is capable of take-off and landing autonomously. But this level of autonomy is minor compared to the spectacular areas where RPA development is being directed.

The shift from RPA to drone: 'autonomous' machines and deskilling

One of the defining characteristics of contemporary RPA platforms like the MQ-1 and MQ-9 is their capability to be deployed for long periods of time as combat or intelligence support. To this end, the 42 AS and 39 squadron (US and UK) are on operations constantly with RPA platforms. Initially, co-habiting at Creech Air Force Base, Nevada, the Royal Air Force had flown over 400 drone missions from October 2007 to 2010 (Webb et al. 2010: 33). Having moved its main base of operations to RAF Waddington in 2012, the UK currently possesses 10 MQ-9 Reapers potentially alleviating the past practice of 'borrowing' American platforms at times of high operational tempo. Pilot and sensor operators are given further support by intelligence specialists, signallers, and meteorologists who help to process the data and images being supplied in real-time on 10 high resolution computer screens. Initially, unlike the United States, only pilots with combat experience could fly RPA platforms in the RAF (Webb et al. 2010: 37). This changed in 2013 when the RAF graduated its first class of 'ground-based' pilots for drone systems. Officials were quick to point out that this new sub-specialisation of personnel would not be entitled to the extra pay given to regular pilots but would remain commissioned officers – contrasting with the United States where non-commissioned personnel do operate drones (Page 2013).

USAF RPA crews for Predators and Reapers can be located at one of several airbases: Cannon, Creech, Davis-Monathan, Holloman, March, Nellis, or Fargo. Systems used outside of the United States and operators who control take-off and landing manoeuvres are located predominantly in eastern Africa, the Middle East, and Central Asia. Tasks are assigned through the Combined Air Operations Centre which is located in Qatar. At Creech, aircrews control Predators and Reapers in real-time via satellite links. Pilots sit in the left hand seat and sensor operators in the right. Flying controls include a throttle and stick while data streams across monitors. Two of these screens provide high resolution video of the battle-space, while the others display information like satellite imagery and communications details (IISS 2009). It has been claimed that operators can use Google Earth on touch screens to point to a location they want to bomb. iPhone and Android apps are being developed for US troops to aid in targeting and communications. Already, one such app called Tactical NAV, which facilitates the sharing of photographs with GPS coordinates, is available for the iPhone.[10]

It is argued that RPA platforms offer levels of inter-unit integration necessary for tactical adaptability – a force multiplying capability championed by the transformation doctrine's promotion of 'jointness' and synergy in netwar. For example, Reaper crews are able to talk directly to Joint Tactical Controllers on the ground that possess the ability to view feeds captured by the RPA on a remote operated video enhanced receiver (ROVER). This provides troops in the field with the same view as can be seen by the crew. MQ-9 crews can also communicate directly with commanders located elsewhere (IISS 2009). These systems play an integral role in linking personnel engaged in the theatre of operations with those who coordinate the strategic systems of command and control in real-time. Thus, in *RM2009* it is argued that:

> each unmanned system becomes a node on the network that contributes to the formation of the network that enables communications and sensor feed flow. Additionally, payloads carried by the unmanned platforms can contribute to deployment of sensors and communications relays that are dedicated to net centric operations, in essence serving as autonomous delivery mechanisms for strategic emplacement of the network communications and sensor components.
>
> (DoD 2009: 15)

As such, the RPA can serve as a direct participant in the swarm or as a system that enables swarm formation with its communications capabilities in the battle-space. The US military remains committed to increasing the levels of automation and autonomy in unmanned systems as appropriate to addressing specific problems confronting 'war-fighters' in specific contexts. To these ends, the *RM2009* proposed that RPA platforms would undergo a process of 'evolutionary adaptation' from 2009 to 2015 in order to increase levels of situational awareness to the point where systems would be able to practice 'dynamic obstacle avoidance'.[11] It was further argued that through a process of 'revolutionary adaptation' planned from 2015 to 2034, RPA platforms would be reconfigured as drones, capable of possessing a level of on-board situational awareness that could conceivably enable them to navigate John Boyd's (1987) 'observation, orientation, decision, and action' loop autonomously. Included amongst the desired autonomous attributes were capabilities such as 'covert and self-concealing behaviours' 'intelligent, adaptive navigation' – operating independently of GPS technology – and biomimetic human detection through visual, sonic, and olfactory sensors. But perhaps most ambitious were the demands for 'bird dog/war-fighter's associate' (BDWA) functionality by 2015 and 'hierarchical collaborative behaviours' (HCB) by 2034. Bird dog capabilities were premised on a system possessing 'empathy with the human operator' so that the platform is able to take 'high-level commands from the operator much as a bird dog does from the hunter'. BDWA requires that the environmental awareness capacity of the unit is expanded to include the 'the operator's mental and physiological status' with cues garnered from readings of human operators to help interpret both verbal and gesture-based commands (DoD 2009: 154). According to the *RM2009*, hierarchical collaborative behaviours are a step beyond BDWA. HCB involve complex task allocation amongst human operators and other systems such that:

> the human commander will be able to control a group of heterogeneous robots through a 'smart' squad leader robot. The robot takes high level plans and goals from the human commander, then formulates the detailed plans, tasks, and monitors other more specialized robots to perform the work.
>
> (DoD 2009: 159)

In retrospect, these desires and the timeframes for their development were hubristic. However, although timelines have been stretched and budgets constrained, RPA doctrine and procurement are being channelled towards the development of extremely complex systems that will require less direct input from human operators. For example, swarms and swarming actions may become self-generating. Moreover, drones are projected to be active multi-taskers, fulfilling multiple roles concurrently. For example, the *RM2009* argued that as unmanned systems become increasingly autonomous, one will see 'on-board sensors that provide the systems with their own organic perception [and these] will be able to contribute to Battle Space Awareness regardless of their intended primary mission' (DoD 2009: 8). Although these levels of total autonomy may be a goal, for the moment, the future development of RPA platforms is still primarily considered as a means for augmenting the flexibility available to 'war-fighters' in completing assigned missions (DoD 2009: 7).

But the current and future use of RPA platforms has raised concerns about the potential deskilling of 'regular' pilots and RPA operators. With the growth in the number of RPA platforms it is less clear what future training regimes will look like, particularly if systems become more autonomous as is planned. Some analysts predict that like the US Army, the USAF may move to offering a reduced amount of training suitable for the RPA but insufficient for other forms of flight. The sense that changes in training programmes and perceptions that RPA platforms are contributing to a deskilling of military pilots raises both issues of service morale but also management challenges over more mundane matters such as how to give credit for flying hours to ground-based pilots (Tirpak 2007: 50). At the same time, it is also recognised that the role of a RPA operator is set to become more complex and far different than flying a regular aircraft as the autonomy of these systems increases. In the not-so-distant future, if the RPA operator becomes less of a pilot of a single aircraft and more of a manager of several drones working in conjunction, it is speculated that there may be effects on how the profession and the technology itself are perceived.

One of the key concerns being raised is if autonomous systems prove viable in testing, will they be able to perform with the same level of ability as a human operator in an actual theatre of operation? This is particularly important with regards to aspects of chaoplexic warfare that stress the value of unpredictability through adaptation. While Stanley McChrystal once boasted that RPAs create conditions in which 'we can have eyes 24/7 on our adversaries', it is how surveillance imagery and intelligence are processed that is more important than the capacity to gather it (Helmore 2009). The IISS (2009) reported that over the course of their three-year tours, RPA crews often become highly attuned to the everyday rhythms of the environment into which they are embedded. It is argued that this knowledge of local patterns of work, travel, and leisure, spatial organisation, and even community membership allows them to discern when something unusual is happening. The feeling that something is 'not quite right' is an intuition that is often established as much by what may be seen as by what is absent from sight within the norms of a given context. The argument then is that autonomous systems are unlikely to be able to approach these levels of situational awareness for some time to come. Thus, a move to autonomous platforms could lead to losses in the ability to pre-emptively discern key changes in the operational environment.

The second set of concerns are the practical implications of legal questions discussed in the second chapter that could arise when autonomous platforms become capable of initiating sequences to kill of their own volition. Ultimately, who would retain control over targeting and weapon-systems? Would these functions become fully automated through targeting algorithms and firing protocols or would these remain under the explicit control of human operators?[12] While many weapons already possess the capability to apply lethal force without operator input (e.g., landmines) such systems have been vilified for applying force indiscriminately. Even if RPA platforms develop the ability to discriminate targets from bystanders – as envisioned – it is difficult to imagine that in a military culture where command and control remains a powerful organising pole and where commanders themselves are adverse to the reckless pursuit of actions that entail legal liabilities, that the full automation of the kill chain sequence when pursuing human beings would be implemented. Thus, it would seem probable that current procedures that provide clearance to fire would remain with intelligence analysts and ground troops playing a role in the process. If this is the case and the need for close human monitoring remains, many of the autonomous capabilities projected for RPA platforms would appear superfluous if not redundant. Yet the desire for autonomous systems may remain strong as their symbolic status in spectacular consumption becomes entrenched.

Desire, spectacle, and RPA proliferation

Within media commentaries, there is an increasing focus on the ways in which RPA platforms (almost uniformly referred to as drones) are proliferating. From their use in police surveillance to their deployment by environmental advocacy groups, the military rationale for their deployment resonates: drones are seen as providing a range of services for a relatively low cost. The potential – of the technology and the market – is considered to be enormous. One European industry analysis report argued that RPA platforms now share similar open-ended possibilities regarding their use as personal computers did in the 1980s (ECEI DG 2007: 62). Online, one can now even find websites run by dedicated hobbyist enthusiasts (e.g., diydrones.com) who design, build, and fly their own drone platforms with varying levels of sophistication – including autonomous capabilities.

The extent to which RPAs become a part of everyday policing, border control, surveillance, monitoring, and other forms of spatial management will in part be determined by existing rules and regulations regarding what is permitted to fly in domestic airspace and who is permitted to pilot them. For example, within the European Union, the RPA is subject to several regulatory bodies including the European Aviation Safety Agency and the European Organisation for Civil Aviation Equipment who need to determine technical standing orders, minimum operational performance standards, and airworthiness criteria (ECEI DG 2007). It is also worth noting that RPA deployment in the European context would need to navigate the normative legacies of international arms control regimes including the Conventional Forces in Europe Treaty (Altmann 2013). Given the increasing tensions between the EU/NATO and Russia, this may pose significant impediments to the development, production, and deployment of RPA platforms in Europe. However, the EU has historically shown a desire to develop and maintain its own defence industry capacity, particularly for emerging markets – like the RPA – that look to be commercially lucrative (e.g., Hayes et al. 2014). The development of a strong RPA industry will necessitate both the economies of scale that civilian use can provide as well as the infrastructure – including airspace – for research and development. While the institutional context may be slightly less byzantine and the levels of political will to push for large-scale deployment may be higher in the United States, there are still myriad technical issues, regulations, and specifications that will need to be determined by the Federal Aviation Authority and Civil Aeronautics Board before RPA platforms could become common-place (US Department of Transport 2013). While the FAA has allowed RPA systems in civilian airspace in the United States – subject to certification – since 2012, the September 2015 deadline for the full integration of RPAs was missed because of 'technical and registry problems' (Whitlock 2014). The number of outstanding issues suggests that full integration may now be more a question of 'if' than 'when' (Whitlock 2014).

In 2012, the FAA released its list of granted authorizations to 63 RPA launch sites in the United States. Licensed operators include the military services, the FBI, Border and Customs patrol, and NASA as well as universities, local police forces, and defence contractors. By November 2015, there had been over 2,300 Section 333 exceptions for drone use granted by the FAA.[13] While there may arguably be a case for why some of organisations require licenses (e.g., universities may be involved in the research and development of drone technologies), for others (e.g., the Ogden Police Department, Utah or the City of Herrington, Kansas) the case is less clear in terms of an obvious need. But the spectacle of the RPA, the conspicuous consumption of drone ownership, and the desire for capabilities that are far in excess of whatever might be needed may play large here (Wall and Monahan 2011; Wall 2013). While targeted killing committed by these agencies within the sovereign territory of the United States is not a realistic future scenario, one can envision their deployment may be motivated by a desire to become a part of an emerging domestic surveillance regime in which drones will begin to play a more prominent role. RPA platforms are represented as the leading edge of security technology with their possession signifying membership in an elite grouping of security forces (or high-tech developers). Michael Salter (2014: 164) has argued that the acquisition of advanced military systems, like RPAs, by the police is catalysed by '...unprecedented social, economic, and cultural change' and reflects a desire to '...embody and enact the militarised subject positions made possible by such technology'. With a range of systems becoming available, demand rising both from within and outside of the military, decommissioning programmes that transfer military equipment to civilian forces, a perception that ownership is conducive to institutionalising effective organisational forms, a perception that platforms are cost effective and task efficient, and with bloated counter-terrorism budgets looking to be spent, and a policing culture that seeks to adopt military operational norms (Campbell and Campbell 2014), the continuing growth of the RPA industry is benefiting from a favourable constellation of politico-economic elements.

Conclusions

This chapter has presented the argument that the recent predominance of RPA technology in war-fighting doctrines goes beyond the processes of techno-logical development that have made increasingly sophisticated systems possible. Rather, the argument has shown how a complex assemblage of science, capitalist cultural values, and war-fighting doctrines, as well as their associated styles of thought and organisational structures, have made possible the emergence of the RPA as a central tool in the irregular warfare said to constitute con-temporary counter-insurgency. In mapping out this constellation of forces enabling RPA development, procurement, and deployment, this chapter has shown how the contributing contingent dynamics have been institutionalised as well as the places where they may meet their limits. While the narrative itself may not have necessarily read this way, the point has been to demonstrate that there has been nothing inevitable about the elevation of the RPA in contemporary counter-insurgency or its contributions to the targeted killing assemblage. More broadly, this chapter has shown that accounting for the scientific, technological, and social elements of war-fighting doctrines – and their associated practices – requires careful consideration of the economic understandings in which these are embedded. By examining these elements of the complex social assemblage constituted by war, one can better expose the broader cultural dimensions of war-fighting practices that might otherwise be overlooked. In the next chapter, the analysis shifts to looking at the aesthetic subjects produced by the targeted killing assemblage and how militaries are problematising them through an arithmetic locus of control.

Notes

1 For example, analyses of procurement patterns indicate that the United States Air Force is acquiring RPA platforms as quickly as it can absorb them.
2 Bousquet (2009: 235) sagely notes that regimes and eras that he identifies should not be seen as distinctly delineated time periods. Rather he argues that they represent '…historical intervals in which the ideas and practices characterised by a particular regime can be seen to be ascendant'.
3 Sennett (2006) illustrates how this logic contained the seeds of its own destruction. Advocates of these rigid top-down models would overlook that any order, no matter how precise, is also subject to interpretation by those who must implement its diktats. He shares stories of how early management and efficiency experts – like Frederick Taylor – were unable to anticipate or adapt to the modulation of directives through interpretation.
4 For example, see Gibson-Graham (1996); Mezzadra (2011); Ritzer and Jurgenson (2010); and Marazzi (2011).
5 On the changing role of the state see Geddes (2005); Miller and Rose (2008); and Peck (2004).
6 For a critical evaluation of these claims, see Form (1987).
7 This is not to argue that mass production did not confer workers with skills. Rather, the argument is that the specialisation and Taylorisation of tasks in industrial manufacturing was not necessarily conducive to the development of a broad and deep skill set in the area of production. Thus, while a classical furniture maker might have been highly skilled in the use of tools to produce a range of furniture items, it was claimed that under industrialisation, a worker might be highly skilled in the use of a particular tool in a narrow part of the manufacturing process.
8 Negative feedback systems are those in which relations between elements contribute to the operation and reproduction of the system. Negative feedback thus works to ensure systemic stability within a closed loop by maintaining equilibrium points. Positive feedback occurs when outputs of a system feed back into the system, leading to systemic change. A slight change in one element – which may be a random occurrence – contributes to increased changes elsewhere in the system. In contra-distinction to negative feedback systems, positive feedback systems are characterised by growth and adaptation, with minimal changes having the potential to be amplified. Thus positive feedback systems have a tendency to become increasingly chaotic with the possibility that changes induced through positive feedback may threaten the survival of the system itself (Bousquet 2009: 166–167).

9 For a historical overview of the development of RMA thinking in the United States, see Rosen (2010).
10 See http://www.tacticalnav.com. Accessed 15 October 2015.
11 While this may seem like a rather mundane capability, its realisation would be premised on giving machines the ability to make complex perceptual distinctions (for example between shadows and actual objects) that would constitute a major technological achievement.
12 For a discussion, see Sparrow (2007); Krishnan (2009); and Sharkey (2010).
13 See http://www.faa.gov/uas/legislative_programs/section_333. Accessed 15 October 2015.

Bibliography

Agger, B. (2004) *Speeding Up Fast Capitalism: Internet Culture, Work, Families, Food, Bodies*, Brookline, Paradigm Publishing, Inc.

Altmann, J. (2013) 'Arms Control for Armed Uninhabited Vehicles: An Ethical Issue', *Ethics and Information Technology*, 15(2), 137–152.

Amin, A. (2008) *Post-Fordism: A Reader*, Oxford, Wiley.

Arnold, M. (2003) 'On the Phenomenology of Technology: the "Janus-Faces" of Mobile Phones', *Information and Organization*, 13(4), 231–256.

Arquilla, J. and Ronfeldt, D. (1993) '"Cyberwar is Coming!"', *Comparative Strategy*, 12(2), 141–165.

Arquilla, J. and Ronfeldt, D. (2000) 'Swarming and the Future of Conflict'. DTIC Document. Available from: http://oai.dtic.mil/oai/oai?verb=getRecord&metadata Prefix=html&identifier=ADA384989. Accessed 15 October 2015.

Balle, J. K. O. (2015) 'About the Predator and Reaper'. *Aeroweb*. Available from:http://www.bga-aeroweb.com/Defense/MQ-1-Predator-MQ-9-Reaper.html. Accessed 15 October 2015.

Baran, P. A. and Sweezy, P. M. (1968) *Monopoly Capital: An Essay on the American Economic and Social Order*, New York, Monthly Review Press.

Barnett, T. P. M. (2005) *The Pentagon's New Map: War and Peace in the Twenty-first Century*, New York, Penguin.

Barrientos, S. W. (2013) '"Labour Chains": Analysing the Role of Labour Contractors in Global Production Networks', *The Journal of Development Studies*, 49(8), 1058–1071.

Baum, J. A. C. and McGahan, A. M. (2013) 'The Reorganization of Legitimate Violence: The Contested Terrain of the Private Military and Security Industry during the Post-Cold War Era', *Research in Organizational Behavior*, 33(1), 3–37.

Beck, U. and Camiller, P. (2000) *What is Globalization?*Cambridge, Polity Press.

Beckett, D. and Hager, P. (2013) *Life, Work and Learning*, Abingdon, Routledge.

Beirne, M. (2013) *Rhetoric and the Politics of Workplace Innovation*, Cheltenham, Edward Elgar Publishing.

Bousquet, A. (2009) *The Scientific Way of Warfare: Order and Chaos on the Battlefields of Modernity*, London, Hurst and Company.

Boyd, J. R. (1987) '"Organic Design for Command and Control", A Discourse on Winning and Losing '. Available from: https://www.quadratic.net/boyd/organic_design.pdf. Accessed 15 October 2015.

Boyle, A. (2012) 'The US and its UAVs: A Cost-Benefit Analysis'. American Security Project. Available from:http://www.americansecurityproject.org/the-us-and-its-uavs-a-cost-benefit-analysis. Accessed 15 October 2015.

Braudel, F. (1981) *Civilization and Capitalism, 15th-18th Century: The Structures of Everyday Life: The Limits of the Possible*, Berkeley and Los Angeles, University of California Press.

Braudel, F. (1982a) *Civilization and Capitalism, 15th-18th Century: The Perspective of the World*, Berkeley and Los Angeles, University of California Press.

Braudel, F. (1982b) *Civilization and Capitalism, 15th-18th Century: The Wheels of Commerce*, Berkeley and Los Angeles, University of California Press.

Campbell, D. (1998) *Writing Security: United States Foreign Policy and the Politics of Identity*, Minneapolis, University of Minnesota Press.

Campbell, D. J. and Campbell, K. M. (2014) 'Police/Military Convergence in the USA as Organisational Mimicry', *Policing and Society* (ahead-of-print), 1–22. DOI: 10.1080/10439463.2014.942852.

Cevik, P., Kocaman, I., Akgul, A. S. and Akca, B. (2013) 'The Small and Silent Force Multiplier: A Swarm UAV – Electronic Attack', *Journal of Intelligent & Robotic Systems*, 70(1–4), 595–608.

Clarke, R. (2014) 'Understanding the Drone Epidemic', *Computer Law & Security Review*, 30(3), 230–246.

Cohen, E. A. (2004) 'Change and Transformation in Military Affairs', *Journal of Strategic Studies*, 27(3), 395–407.

Collinson, D. L. and Collinson, M. (1997) '"Delayering Managers": Time-Space Surveillance and its Gendered Effects', *Organization*, 4(3), 375–407.

Czarnitzki, D., Grimpe, C. and Toole, A. A. (2014) 'Delay and Secrecy: Does Industry Sponsorship Jeopardize Disclosure of Academic Research?', *Industrial and Corporate Change*, 24(1), 251–279.

Davidow, B. (2012) 'How Computers Are Creating a Second Economy Without Workers'. *The Atlantic*. 10 April. Available from: http://www.theatlantic.com/busi ness/archive/2012/04/how-computers-are-creating-a-second-economy-without-workers/ 255618. Accessed 15 October 2015.

Davies, T. (2012) 'How Might Open Data Contribute to Good Governance?', *Commonwealth Governance Handbook*, 148–150. Available from: http://www.comm onwealthgovernance.org/assets/uploads/2012/10/How-might-open-data-contribute-to-good-governance.pdf. Accessed 15 October 2015.

Davies, T. G. and Bawa, Z. A. (2012) 'The Promises and Perils of Open Government Data (OGD)', *The Journal of Community Informatics*, 8(2). Available from: http:// www.ci-journal.net/index.php/ciej/article/view/929/955. Accessed 15 October 2015.

Debord, G. ([1967] 1995) *The Society of the Spectacle*, New York, Zone Books.

Deitelhoff, N. and Geis, A. (2009) 'Securing the State, Undermining Democracy: Internationalization and Privatization of Western Militaries'. Universitat Bremen. Available from: http://www.econstor.eu/handle/10419/27914. Accessed 15 October 2015.

Dillon, M. and Reid, J. (2009) *The Liberal Way of Warfare: Killing to Make Life Live*, London, Routledge.

Doogan, K. (2001) 'Insecurity and Long-term Employment', *Work, Employment & Society*, 15(3), 419–441.

Duffield, M. (2001) *Global Governance and the New Wars: The Merging of Development and Security*, London, Zed Books.

Dunn, D. H. (2013) 'Drones: Disembodied Aerial Warfare and the Unarticulated Threat', *International Affairs*, 89(5), 1237–1246.

Edwards, C. (2007) 'The Automatic Airforce', *Engineering and Technology*, 2(4), 32–35.

[ECEI DG] European Commission Enterprise and Industry Directorate General (2007) 'Study Analysing the Current Activities in the Field of UAV'. *ECEI DG.* Available from: http://ec.europa.eu/enterprise/policies/security/files/uav_study_elem ent_2_en.pdf. Accessed 15 October 2015.

Fearfull, A. and Dowling, M. (2011) 'Skills in the Twenty-first-century Organization: The Career of a Notion' in K. Townsend and A. Wilkinson (eds), *Research Handbook on the Future of Work and Employment Relations.* Cheltenham, Edgar Elgar, 169–187.

Fellman, P. V. and Wright, R. (2004) 'Modeling Terrorist Networks, Complex Systems at the Mid-range', arXiv preprint arXiv:1405.6989.

Feyerabend, P. K. (1999) *Knowledge, Science and Relativism*, Cambridge, Cambridge University Press.

Fiorani, F. (2005) *The Marvel of Maps: Art, Cartography, and Politics in Renaissance Italy*, New Haven, Yale University Press.

Form, W. (1987) 'On the Degradation of Skills', *Annual Review of Sociology*, 13(1), 29–47.

Foucault, M. (1977) *Discipline and Punish: The Birth of the Prison*, New York, Pantheon Books.

Geddes, M. (2005) 'Neoliberalism and Local Governance – Cross-national Perspectives and Speculations', *Policy Studies*, 26(3–4), 359–377.

Gertler, J. (2012) *US Unmanned Aerial Systems.* Washington, DC: Library of Congress, Congressional Research Service. Available from: http://www.dtic.mil/cgi-bin/ GetTRDoc?AD=ADA566235. Accessed 15 October 2015.

Gibson-Graham, J. K. (1996) *The End of Capitalism (As We Knew It): A Feminist Critique of Political Economy*, Minneapolis: University of Minnesota Press.

Gill, S. (1995) 'Globalisation, Market Civilisation, and Disciplinary Neoliberalism', *Millennium*, 24(3), 399–423.

Gill, S. (1998) 'New Constitutionalism, Democratisation and Global Political Economy', *Pacifica Review*, 10(1), 23–38.

Godfrey, R., Brewis, J., Grady, J. and Grocott, C. (2014) 'The Private Military Industry and Neoliberal Imperialism: Mapping the Terrain', *Organization*, 21(1), 106–125.

Gregory, D. (2004) *The Colonial Present*, Oxford, Blackwell Publishing.

Gregory, D. (Forthcoming) 'Moving Targets and Violent Cartographies'. *Geographical Imaginations.* Available from: http://geographicalimaginations.files.wordpress.com/ 2012/07/derek-gregory-moving-targets-and-violent-geographies-final.pdf. Accessed 15 October 2015.

Guha, M. (2011) *Reimagining War in the 21st Century: From Clausewitz to Network Centric Warfare*, Abingdon, Routledge.

Hastings-Dunn, D. (2013) 'Drones: Disembodied Aerial Warfare and the Unarticulated Threat', *International Affairs*, 89(5), 1237–1246.

Hall, A. R. and Coyne, C. J. (2014) 'The Political Economy of Drones', *Defence and Peace Economics*, 25(5), 445–460.

Harcourt, B. E. (2014) 'Governing, Exchanging, Securing: Big Data and the Production of Digital Knowledge', Columbia Public Law Research Paper (14–390). Available from: http://papers.ssrn.com/sol3/papers.cfm?abstract_id=2443515. Accessed 15 October 2015.

Hardt, M. and Negri, A. (2000) *Empire*, Cambridge, Harvard University Press.

Harvey, D. (1989) *The Condition of Postmodernity: An Enquiry into the Politics of Cultural Change*, New York, Oxford University Press.

Hassan, R. (2009) *Empires of Speed: Time and the Acceleration of Politics and Society*, Leiden, Brill.

Hassard, J., Morris, J. and McCann, L. (2012) '"My Brilliant Career"? New Organizational Forms and Changing Managerial Careers in Japan, the UK, and USA', *Journal of Management Studies*, 49(3), 571–599.

Hayes, B., Jones, C. and Topfer, E. (2014) *Eurodrones Inc.* Amsterdam: StateWatch. Available from: http://www.statewatch.org/news/2014/feb/sw-tni-eurodrones-inc-feb-2014.pdf. Accessed 15 October 2015.

Helmore, E. (2009) 'US Air Force Prepares Drones to End Era of Fighter Pilots'. *Guardian*. 23 August. Available from: http://www.theguardian.com/world/2009/aug/22/us-air-force-drones-pilots-afghanistan. Accessed 15 October 2015.

Henkin, Y. (2014) 'On Swarming: Success and Failure in Multidirectional Warfare: from Normandy to the Second Lebanon War', *Defence Studies*, 14(3), 310–332.

Hobson, J. A. (1938) *Imperialism: A Study*, Nottingham, Spokesman Books.

Hottenrott, H. and Lawson, C. (2014) 'Research Grants, Sources of Ideas and the Effects on Academic Research', *Economics of Innovation and New Technology*, 23(2), 109–133.

Hsu, K., Murray, C., Cook, J. and Feld, A. (2013) *China's Military Unmanned Aerial Vehicle Industry*, Washington, DC: US-China Economic and Security Review Commission.

Huff, T. E. (2003) *The Rise of Early Modern Science: Islam, China, and the West*, Cambridge, Cambridge University Press.

Hughes, A. and Reimer, S. (eds) (2004) *Geographies of Commodity Chains*, London, Routledge.

Ilachinski, A. (2012) 'Modelling Insurgent and Terrorist Networks as Self-organised Complex Adaptive Systems', *International Journal of Parallel, Emergent and Distributed Systems*, 27(1), 45–77.

(IISS) International Institute for Strategic Studies (2009) 'The Drones of War', *Strategic Comments*, 15(4), 1–2.

Irvin, D. W. (2003) *History of Strategic Drone Operations*, Nashville, Turner Publishing Company.

Isenberg, D. (2007) *Budgeting for Empire: The Effect of Iraq and Afghanistan on Military Forces, Budgets, and Plans*, Oakland, Independent Institute.

Jay, M. (1988) 'Scopic Regimes of Modernity', *Vision and Visuality*, 2(1), 3–38.

Jensen, P. H. and Webster, E. (2011) 'Do Patents Influence Academic Scientists' Choice of Research Projects?', *University of Melbourne Working Paper*. Available from: ftp://193.196.11.222/pub/zewdocs/veranstaltungen/innovationpatenting2011/papers/Jensen.pdf. Accessed 15 October 2015.

Johnson, C. (2003) 'The War Business: Squeezing Profits from the Wreckage in Iraq'. *Harper's Magazine*. 307(1842), 53–58.

Johnston, J. M. and Girth, A. M. (2012) 'Government Contracts and "Managing the Market" Exploring the Costs of Strategic Management Responses to Weak Vendor Competition', *Administration & Society*, 44(1), 3–29.

Kalleberg, A. L. (2011) *Good Jobs, Bad Jobs: The Rise of Polarized and Precarious Employment Systems in the United States, 1970s-2000s*, New York, Russell Sage Foundation.

Kautsky, K. (2007) *Ultra-imperialism*, Marlborough, Adam Matthew Digital.

Kennedy, P. (2010) *The Rise and Fall of the Great Powers*, New York, Random House Digital, Inc.

Kirpal, S. (2011) *Labour-market Flexibility and Individual Careers: A Comparative Study*, New York, Springer.

Kitchin, R. and Dodge, M. (2011) *Code/space: Software and Everyday Life*, Cambridge, MIT Press.

Klein, N. (2007) *The Shock Doctrine: The Rise of Disaster Capitalism*, London, Macmillan.

Kreps, S. and Zenko, M. (2014) 'The Next Drone Wars; Preparing for Proliferation', *Foreign Affairs*, 93(1), 68–79.

Krishnan, A. (2008) *War as Business: Technological Change and Military Service Contracting*, Aldershot, Ashgate Publishing, Ltd.

Krishnan, A. (2009) *Killer Robots: Legality and Ethicality of Autonomous Weapons*, Aldershot, Ashgate Publishing, Ltd.

Kuhn, T. S. (2012) *The Structure of Scientific Revolutions*, Chicago, University of Chicago Press.

Lee, C. K. and Kofman, Y. (2012) 'The Politics of Precarity Views beyond the United States', *Work and Occupations*, 39(4), 388–408.

Lenin, V. I. (1999) *Imperialism: The Highest Stage of Capitalism*, Broadway, Resistance Books.

Levy, D. L. (2005) 'Offshoring in the New Global Political Economy', *Journal of Management Studies*, 42(3), 685–693.

Levy, F. and Murnane, R. J. (2004) *The New Division of Labor: How Computers are Creating the Next Job Market*, Princeton, Princeton University Press.

Madrick, J. (2012) 'The Deliberate Low-wage, High-insecurity Economic Model', *Work and Occupations*, 39(4), 321–330.

Marazzi, C. (2011) *The Violence of Financial Capitalism*, Cambridge, MIT Press Books.

Medina, D. A. (2014) 'Drone Markets Open in Russia, China and Rogue States as America's Wars Wane'. *Guardian*. 22 June. Available from: http://www.theguardian.com/business/2014/jun/22/drones-market-us-military-china-russia-rogue-state. Accessed 15 October 2015.

Meiter, J. S. (2006) 'Network Enabled Capability: A Theory Desperately in Need of Doctrine', *Defence Studies*, 6(2), 189–214.

Melman, S. (1970) *Pentagon Capitalism: The Political Economy of War*, Columbus, McGraw-Hill.

Merton, R. K. (1973) *The Sociology of Science*, Chicago, University of Chicago Press.

Metz, S. (2007) 'America's Defense Transformation: A Conceptual and Political History', *Defence Studies*, 6(1), 1–25.

Mezzadra, S. (2011) 'How Many Histories of Labour? Towards a Theory of Postcolonial Capitalism', *Postcolonial Studies*, 14(2), 151–170.

Miller, P. and Rose, N. (2008) *Governing the Present*, Cambridge, Polity Press.

Mills, C. W. (1956) *The Power Elite*, Oxford, Oxford University Press.

Mitchell, P. T. (2013) *Network Centric Warfare: Coalition Operations in the Age of US Military Primacy*, Abingdon, Routledge.

Moffat, J. (2010) *Complexity Theory and Network Centric Warfare*, Collingdale, Diane Publishing.

Nitzan, J. and Bichler, S. (1995) 'Bringing Capital Accumulation Back in: The Weapondollar-Petrodollar Coalition − Military Contractors, Oil Companies and Middle East "Energy Conflicts"', *Review of Political Economy*, 2(3), 446–515.

Nolin, P. C. (2012) 'Unmanned Aerial Vehicles: Opportunities and Challenges for the Alliance'. *NATO*. Available from: http://5ton.pl/wp-content/uploads/2015/02/20121120_drones_report_natopa.pdf. Accessed 15 October 2015.

Nowotny, H., Scott, P. and Gibbons, M. T. (2013) *Re-thinking Science: Knowledge and the Public in an Age of Uncertainty*, Cambridge, Polity.

OECD (2010) *Main Science and Technology Indicators 2010*. Paris, OECD.

Page, L. (2013) 'RAF Graduates First Class of Groundbased "Pilots"'. *The Register*. Available from: http://www.theregister.co.uk/2013/04/04/raf_drone_rpas_pilots_graduate. Accessed 15 October 2015.

Pang, C. K., Le, C. V., Gan, O. P., Hudas, G., Middleton, M. B. and Lewis, F. L. (2012) 'Discrete Event Command and Control of Multiple Military Missions in Network Centric Warfare'. *Proc. of the 8th International Conference on Intelligent Unmanned Systems (ICIUS 2012)*, 74–79.

Peck, J. (2004) 'Geography and Public Policy: Constructions of Neoliberalism', *Progress in Human Geography*, 28(3), 392–405.

Polanyi, K. (1944) *The Great Transformation*, New York, Rinehart.

Popper, K. (2011) *The Open Society and its Enemies*, London, Routledge.

Pyper, D. and McGuiness, F. (2014) 'Zero-hours contracts'. *Business and Transport Section.* London: House of Commons Library.

Reilly, P. A. (1998) 'Balancing Flexibility – Meeting the Interests of Employer and Employee', *European Journal of Work and Organizational Psychology*, 7(1), 7–22.

Reno, W. (1998) *Warlord Politics and African States*, Boulder, Lynne Rienner Publishers.

Ritzer, G. and Jurgenson, N. (2010) 'Production, Consumption, Prosumption: The Nature of Capitalism in the Age of the Digital "Prosumer"', *Journal of Consumer Culture*, 10(1), 13–36.

Robin, C. (2004) *Fear: The History of a Political Idea*, Oxford, Oxford University Press.

Rosen, S. P. (2010) 'The Impact of the Office of Net Assessment on the American Military in the Matter of the Revolution in Military Affairs', *Journal of Strategic Studies*, 33(4), 469–482.

Ross, A. (ed.) (1996) *Science Wars*, Durham, Duke University Press.

Ross, A. K. (2013) 'Erased US Data Shows 1 in 4 Missiles in Afghan Airstrikes Now Fired by Drone. *Bureau for Investigative Journalism*. Available from: http://www.thebureauinvestigates.com/2013/03/12/erased-us-data-shows-1-in-4-missiles-in-afghan-airstrikes-now-fired-by-drone. Accessed 15 October 2015.

Ross, A. K. (2014) 'UK's New Reaper Drones Remain Grounded, Months before Afghan Withdrawal'. *Bureau for Investigative Journalism*. Available from: http://www.thebureauinvestigates.com/2014/05/22/uks-new-reaper-drones-remain-grounded-months-before-afghan-withdrawal. Accessed 15 October 2015.

Rossi, A., Luinstra, A. and Pickles, J. (eds) (2014) *Towards Better Work: Understanding Labour in Apparel Global Value Chains*, New York: Palgrave Macmillan.

Rumsfeld, D. H. (2002) 'Transforming the Military', *Foreign Affairs*, 81(3), 20–32.

Ryan, M. (2014) '"Full Spectrum Dominance": Donald Rumsfeld, the Department of Defense, and US Irregular Warfare Strategy, 2001–2008', *Small Wars & Insurgencies*, 25(1), 41–68.

Salter, M. (2014) 'Toys for the Boys? Drones, Pleasure and Popular Culture in the Militarisation of Policing', *Critical Criminology*, 22(2), 163–177.

Schwartz, M. and Church, J. (2013) 'Department of Defense's Use of Contractors to Support Military Operations: Background, Analysis, and Issues for Congress'.

DTIC Document. Available from: http://oai.dtic.mil/oai/oai?verb=getRecord&meta dataPrefix=html&identifier=ADA590715. Accessed 15 October 2015.

Sengenberger, W. (2005) *Globalization and Social Progress: The Role and Impact of International Labour Standards*, Bonn, Friedrich-Ebert-Stiftung.

Sennett, R. (2006) *The Culture of the New Capitalism*, New Haven, Yale University Press.

Sharkey, N. (2010) 'Saying "No!"to Lethal Autonomous Targeting', *Journal of Military Ethics*, 9(4), 369–383.

Singer, P. W. (2009) *Wired for War: The Robotics Revolution and Conflict in the Twenty-first century*, New York, Penguin.

Sloan, E. C. (2012) *Modern Military Strategy: An Introduction*, New York, Routledge.

Smith, S. (2010) 'US Drones: Inside the Tools of Modern Warfare'. *Channel 4*. Available from: http://www.channel4.com/news/articles/world/americas/us+drones+inside+the+tools+of+modern+warfare/3672737.html. Accessed 15 October 2015.

Sparrow, R. (2007) 'Killer Robots', *Journal of Applied Philosophy*, 24(1), 62–77.

Spohrer, J., Golinelli, G. M., Piciocchi, P. and Bassano, C. (2010) 'An Integrated SS-VSA Analysis of Changing Job Roles', *Service Science*, 2(1–2), 1–20.

Stanger, A. (2009) *One Nation under Contract: The Outsourcing of American Power and the Future of Foreign Policy*, New Haven, Yale University Press.

Stiglitz, J. E. (2000) 'Capital Market Liberalization, Economic Growth, and Instability', *World Development*, 28(6), 1075–1086.

Sutherland, T. (2014) 'Getting Nowhere Fast: A Teleological Conception of Socio-technical Acceleration', *Time & Society*, 23(1), 49–68.

Thompson, P. (2013) 'Financialization and the Workplace: Extending and Applying the Disconnected Capitalism Thesis', *Work, Employment & Society*, 27(3), 472–488.

Tirpak, J. A. (2007) 'UAVs With Bite', *Air Force Magazine* (January), 46–50.

United States Air Force (2007) *Irregular Warfare: Air Force Doctrine Document 2–3*, United States Air Force.

United States Department of Defense (2009) *Unmanned Systems Integrated Roadmap: FY 2009-2034*. Washington, DC: Department of Defense.

United States Department of Defense (2013) *Unmanned Systems Integrated Roadmap: FY2013-2038*. Washington, DC: Department of Defense.

United States Department of Transport (2013) *Integration of Civil Unmanned Aircraft Systems (UAS) in the National Airspace System (NAS) Roadmap*. Washington, DC: US Department of Transport, Federal Aviation Industry.

Virilio, P. (2005) *Negative Horizon*, London, Continuum.

Vostal, F. (2014) 'Thematizing Speed: Between Critical Theory and Cultural Analysis', *European Journal of Social Theory*, 17(1), 95–114.

Wall, T. (2013) 'Unmanning the Police Manhunt: Vertical Security as Pacification', *Socialist Studies/Études socialistes*, 9(2), 32–56.

Wall, T. and Monahan, T. (2011) 'Surveillance and Violence from Afar: The Politics of Drones and Liminal Security-scapes', *Theoretical Criminology*, 15(3), 239–254.

Webb, D., Wirbel, L. and Sulzman, B. (2010) 'From Space, No One Can Watch You Die', *Peace Review*, 22(1), 31–39.

Wheeler, W. (2012) 'The MQ-9's Cost and Performance'. *Time Magazine*. 28 February. Available from: http://nation.time.com/2012/02/28/2-the-mq-9s-cost-and-performance. Accessed 15 October 2015.

Whitlock, C. (2014) 'FAA Will Miss Deadline to Integrate Drones in U.S. Skies, Report Says'. *Washington Post*. 30 June. Available from: http://www.washingtonpost.com/

world/national-security/faa-will-miss-deadline-to-integrate-drones-in-us-skies-report-sa ys/2014/06/30/fd58e8e2–007f-11e4-b8ff-89afd3fad6bd_story.html. Accessed 15 October 2015.

Williams, A. J. (2011) 'Enabling Persistent Presence? Performing the Embodied Geopolitics of Unmanned Aerial Vehicle Assemblage', *Political Geography*, 30(7), 381–390.

Williamson, J. (1993) 'Democracy and the "Washington Consensus"', *World Development*, 21(8), 1329–1336.

Wilson, J. R. (2002) 'UAVs and the Human Factor', *Aerospace America*, 40(7), 53–57.

Wilthagen, T. and Tros, F. (2004) 'The Concept of "Flexicurity": A New Approach to Regulating Employment and Labour Markets', *Transfer: European Review of Labour and Research*, 10(2), 166–186.

Woods, C. and Ross, A. K. (2012) 'Revealed: US and Britain Launched 1,200 Drone Strikes in Recent Wars. *Bureau for Investigative Journalism*. Available from: http:// www.thebureauinvestigates.com/2012/12/04/revealed-us-and-britain-launched-1200-drone-strikes-in-recent-wars. Accessed 15 October 2015.

5 The aesthetic subjects of targeted killing

Introduction

In the previous three chapters, the law, narratives of violence, and relationships connecting military strategy, capitalist values, and scientific thought have been examined. These have been identified as key elements and variable processes of the targeted killing assemblage that contribute to its conditions of possibility. It has been argued that they have produced problematisations that are central to its territorialisation, that is, the processes through which the targeted killing assemblage is able to reproduce itself. At the same time, it has been recognised that there has been nothing inevitable in these outcomes and that their processes of coding and recoding reveal a cultural politics at work that can be both direct and diffuse. In this chapter, the analysis asks 'what kinds of subjects are produced through the targeting killing assemblage when the deployment of violence is modulated through problematisations focused on control?' Or, to articulate it differently, what kinds of subjects are produced through the excesses of the targeted killing assemblage? To answer such questions requires that one move beyond orthodox understandings of subjectivity that have been predominant in the social sciences.

In *The Time of the City: Politics, Philosophy, and Genre,* Michael Shapiro (2010) makes an important distinction between statistical subjects and aesthetic subjects. For Shapiro, the statistical subject is the disembodied, calculable, *homo oeconomicus* that is both a guiding assumption for – and an aggregate object rendered by – orthodox social scientific research. In contrast, aesthetic subjects are individuals whose everyday experiences reveal the power-relations and socio-cultural dynamics at work in their places of residence, labour, worship, transit, care, and leisure. Moreover, the aesthetic subject helps to expose the multiple identities historically produced within and through these locales. Shapiro (2010: 7) argues that unlike the statistical subject of orthodox social science who blurs 'the ideational fault-lines' that become evident as subjects navigate the places of everyday life and those who co-habit within them, the aesthetic subject reveals their complexity. It is from these interactions that socio-cultural dynamics are produced, experienced, and felt. Thus for Shapiro (2010: 7):

> Aesthetic subjects cannot be gathered arithmetically because their role is not to reflect individual attitudes but to enact the complex political

habitus within which they strive to manage responsibilities, to flourish, or merely to survive.

Most important, aesthetic subjects are not innate subject positions. Rather, they reveal themselves through forms of analysis that aim to disrupt orthodox understandings of the world common to social science research. By attempting to be disruptive of current common-sense understandings, this chapter seeks to explore some of the aesthetic subjects that are produced by the practice of targeted killing and the specific processes through which drone warfare contributes to their emergence. Examining the aesthetic and affective dimensions of targeted killing allows one to move beyond the technocratic rationalisations that pervade orthodox social science as well as critical legal and governmental approaches that seek to challenge these practices. A focus on aesthetic subjects illustrates that the practices of modern high tech network-centric warfare are imbricated in a series of cultural practices – a set of practices that too often escape reflection. When reintroduced to critical scrutiny these practices may become strange, problematic, and difficult once again.

The analysis that follows focuses on three of the aesthetic subjects produced through the practices of targeted killing: remotely piloted aircraft (RPA) operators, those upon whom RPA strikes are initiated, and the RPA itself as an interface and conduit for 'distributions of the sensible' (Rancière 2010: 36–37). Subjectivity and its affective dimensions are explored through the practices of targeted killing and the technologies that enable drone warfare.[1] Of particular interest are the ways in which subjectivity and affect are co-produced and circulated through these domains. Affects can be distinguished from feelings and emotions. Eric Shouse (2005) has argued that 'feelings are *personal,* emotions are *social,* and affects are *prepersonal'*. Affects are non-conscious experiences of (physiological) intensity that are pre-discursive. Feelings are comprised of sensations that we subjectively interpret and define. Emotions are the outward display of feelings – implying a social or inter-subjective element to them. The relationship between affect and feeling is complicated; however, Shouse (2005) argues it can be simplified as follows:

> Without affect feelings do not 'feel' because they have no intensity… In short, affect plays an important role in determining the relationship between our bodies, our environment, and others, and the subjective experience that we feel/think as affect dissolves into experience.

While there is a tendency within political analysis to privilege human agents in the examination of politics, the analysis below demonstrates that the RPA itself is a key nodal point for producing affective, emotional and subjective dynamics that are felt. In this way, RPAs play a key role in the distribution of the sensible that governs contemporary counter-insurgency; that is, by 'establishing the modes of perception' RPA warfare produces 'the places and forms of participation' within which counter-insurgency is inscribed (Rancière

2006: 85). In doing so, as suggested by Rancière (2006: 13), it engages with a (geo)politics that, '...revolves around what is seen and what can be said about it, around who has the ability to see and the talent to speak, around the properties of space and possibilities of time'.

Currently, the stakes in RPA warfare are particularly high. Ultimately, as has been shown in previous chapters, it is a key forum in which decisions about who, where, when, and how to kill are deliberated and enacted. Moreover, it is a forum in which some subjects – and not others – must live within a social environment in which the RPA and its destructive capability are ever present. The effects on subjectivity can be significant. As Joanna Bourke has argued, proximity to killing can have '...a spiritual resonance and an aesthetic poignancy...' for the subjects that it produces (Bourke 1999: 14).

But the aesthetic subjects produced by targeted killing via the RPA do not arise solely from the technical rationality of network centric warfare or counter-insurgency. This chapter demonstrates that:

> intimate acts of killing in war are committed by historical subjects imbued with language, emotion and desire. Killing in wartime is inseparable from wider social and cultural concerns. Combat does not terminate social relationships: rather, it restructures them.
>
> (Bourke 1999: 12)

Part of the interest in restructurings enabled by intimate killing is how attempts to cast RPAs, RPA operators, and insurgents as *homo oeconomicus* are always undermined by affective contingencies, producing new power-relations and social relationships that fall outside of traditional arithmetical notions of subject, object, and agency.

To label these acts of targeted killing via the drone as intimate is not an uncontroversial move. Many critics have argued that the RPA depersonalises killing, making it more convenient by fostering a 'playstation mentality' (Alston 2010: 25; Cole et al. 2010). This characterisation unfortunately mis-characterises the targeted killing assemblage and the ways it is informed by ways of seeing, ways of being, and ways of understanding in the quotidian spaces of the 'everywhere war' (Gregory 2011a). For example, as will be shown below, evidence suggests that RPA pilots often become well-acquainted with those that they target for elimination through long periods of surveillance and monitoring. They often become familiar with the routines, habits, and social circles of those they kill as well as those forces they believe themselves to be protecting. As Derek Gregory (20011b) has suggested, far from being detached, RPA operators are highly embedded – albeit in particular ways – into the environments in which they operate. But the particularities of these embedded forms – and their limits – are also extremely important. So too are the material properties of RPAs and how these contribute to forms of affective embedding through the transmission of flows of information and sensations that may (or may not) become embodied in the operator.

M. Arnold (2003: 232) has argued that 'technologies are not simply a mechanism for achieving a given outcome, where desires, means, and ends can be understood in reasonably unambiguous, linear, and stable terms'. Rather, it is important to examine how technology '…reconstitutes desires and ends, as well as mechanisms, and to account for this reconstituted sociotechnical landscape' (Arnold 2003: 232). Technology 'operates at a metaphysical level, and is not simply instrumental' and thus a platform like the RPA is both produced, and productive of, the cultural, strategic, and affective frames that situate war and political violence (Arnold 2003: 239). At the same time, in attempting to sketch out how particular aesthetic subjects are produced requires an acknowledgement that this is never unidirectional, linear, directly causal, or even necessarily intentional; 'contrariness, paradox, and irony…[may] arise within the analytic frame' (Arnold 2003: 232). Thus, while the RPA may be the interface through which particular aesthetic, affective, and power-relations are linked, the parameters that define these relations at any given time are subject to change.

It must also be noted that the conclusions reached here on the aesthetic subjects of targeted killing can only be suggestive and partial given the need for intensive research in locations and amongst communities that have been relatively closed to outside investigation (e.g., RPA squadrons) or that have traditionally been seen in grand narratives of international relations as unimportant to understanding the relations of power underpinning contemporary geopolitics (e.g., those who live in the 'federally administered tribal areas' along the Afghanistan–Pakistan border). This is necessarily an incomplete account, bounded by my own cultural, linguistic, and national horizons. But the hope is that the preliminary analysis undertaken here on aesthetic subjects can serve as a catalyst for asking a different set of questions about targeted killing and where analysts look for answers to them.

The chapter proceeds as follows: First, the RPA itself is examined as an aesthetic subject. It is argued that problematisations surrounding the control of the RPA reveal how its materiality is being gendered. Second, the examination of aesthetic subjects is extended to RPA operators. The analysis shows how the commissioning of targeted killing via the drone produces embodied affects that are culturally mediated as well as how militaries are seeking to control these through myriad means. Third, populations of interest are read aesthetically to emphasise forms of agency that can undermine the linear cause and effect understandings of how targeted killing works to produce desired outcomes. The chapter concludes with the argument that contra the assumptions of military authorities, target killing involves affective and embodied experiences that cannot be controlled.

The RPA as aesthetic subject

Traditionally, socio-political analysis has focused on individuals, groups, and institutions – with the latter two being considered transposed individuals in aggregate – as exercising agency, that is being able to shape, reshape, contest, acquiesce to, or define the relations of power that constitute any given society. Recently, the predominance of the human (or aggregations of humans) to socio-political analysis has come into question. Whether through Bruno Latour's actor-network theory, forms of object oriented ontology, or agential realist accounts, humans (as subjects), objects, and the spaces in which they operate are conceptualised as an ontologically equal ensemble (i.e., all are accorded similar value as actors that shape phenomena from first principles, subject to actual practices undertaken within a specific context) (e.g., Barad 2003; Bennett 2005; Latour 2005). This materialist (re)turn, for example evidenced in the work of Jane Bennett (2004) and Timothy Mitchell (2002) amongst others, argues that what we may have once dismissed as objects without agency should in fact be seen as agential – or in Latour's terminology as actants – exhibiting forms of behaviour – including resistance – that extend beyond what could be inferred from their design, intended function, capacities, capabilities, or programming. As Mitchell (2002: 38) argues, natural and material forces that are subject to human knowledge and intervention always produce '…certain effects that…[go] beyond…calculations, certain forces that…[exceed] human intention'. The problematique then becomes determining the agency of the technological and natural worlds and the ways in which they produce political action. It also requires that material and/or embodied linkages amongst elements are subject to exploration. A question for analysis then becomes, to paraphrase Mitchell, can the drone speak?

But while agency, its locations, and ontological status are important philosophical – and potentially politically pregnant – questions, they are not a primary concern here.[2] Outside of philosophy, these questions have been a motivating factor in a range of work within critical security studies that has sought to draw attention to how things actively shape those effects that are said to constitute the security field itself (Aradau 2010; Coward 2006; 2009; Peoples 2009; Walters 2014). Thus, at the heart of a (re)turn to the material is an acknowledgement of the ways that material objects, semiotic structures, and agency contribute to the complex ways that (in)security and violence are produced, felt, understood, and experienced. RPAs are illustrative of this complexity, demonstrating the material, discursive, and semiotic links underpinning the practice of contemporary counter-insurgency warfare. They speak to Jane Bennett's (2005: 463) contention that '…the productive power behind effects is always a collectivity' as well as Caroline Holmqvist's (2013: 563) call to pay greater attention to '…the materiality of sentience and…the humanlike characteristics of machines'. Thus, the analysis that follows pursues two parallel lines. The first is to examine how the discursive field in which the RPA is located invests what is already known about the technological properties and design of these platforms into representations of the RPA as a political subject with agential qualities. The second line of examination is to investigate that which cannot be captured or exceeds the expectations about the realm of possible agential qualities of the RPA. It is in this line that notions of autonomy become important to the RPA as an aesthetic subject. While the known capabilities of current platforms fall well below levels of autonomy desired in future systems as outlined in the last chapter, incidents have taken place in which the limits of these capabilities have been exceeded. The unknown and the unanticipated, whether understood as a malfunction or resistance, complicate any certainty about the agential qualities that are held about the RPA. Thus, the ways in which agency and subjectivity are attributed and/or denied reveal power-relations and socio-cultural dynamics at work in understanding the RPA as an aesthetic subject. At the same time, it is important not to neglect the material factors that contribute to this subjectivity.

To begin, it is important to map out how the RPA is constructed as a subject in what Carol Cohn (1987) has referred to as 'techno-strategic discourse'. Within techno-strategic discourse, the RPA is assumed to possess certain qualities and characteristics that are central to the construction of a *habitus* (i.e., socialised norms that shape behaviours) and a field (i.e., counter-insurgency) through which targeted killing takes place. Thus in these types of discussions, the RPA undergoes processes of technological auditing and anthropomorphism. These processes produce a profile filled with accolades and with contradictions. RPAs are presented as both technologically advanced and technologically limited. But as Cohn's work on nuclear weapons has previously indicated, the accolades, gripes, and distinctions made within this dichotomy are highly gendered: RPAs are concurrently presented as persistent, vigilant, protective, sexually potent, temperamental, nurturing, and even recalcitrant.

Proponents and opponents often share a common ground in terms of the tactical impact of RPAs. Similarly, within popular discussions in the press, the RPA is often understood as an object that initially arises independently of any particular socio-political context. It is primarily a technological triumph. Upon its emergence, the RPA is then described as a unit of technology that single-handedly changes the face of modern warfare through its innate capabilities (Bergen and Rothenberg 2014; Editors of Scientific American 2013). However, the RPA, as has been shown in previous chapters, is deeply imbricated in a political-economic and legal-strategic context. And within this context, the RPA is positioned as a sensory system within a broader system of sensors that form a circuit or node within a vast global communications network (e.g., Gregory 2015). This network includes phone lines, satellites, intra-webs, radio waves, ground forces, military commanders, and other vehicles with sensory capabilities. For example, for a RPA in Afghanistan to be controlled in flight by a pilot in the United States involves phone lines carrying flight commands from the United States to a satellite dish in Europe. This satellite then must transmit signals to a small receiving dish on the RPA via the Ku-band satellite data link for beyond line-of-sight operations, that is, for flying at distance beyond normal radio-wave communication (Martin and Sasser 2010: 22; see also McCurley and Maurer 2015: 27). Over the same channel, the RPA sends back three streams of visual content (Martin and Sasser 2010: 30). The supposed accuracy of these sensory systems (e.g., the range at which they can discern objects and movement and the various spectrums at which they can sense including thermal energy) provide a wealth of information that must then be interpreted, both by RPA operators but also by intelligence officers at arm's length who have access to a broader archive of battle-space information. At Langley Air Force Base in Virginia, the monitors showing live feeds from drones flying over Afghanistan are referred to as 'Death TV' (Benjamin 2012: location 116). Intelligence analysts stationed at Langley collect, collate, and review an archive of human intelligence, photos, and signal intelligence, communicating the latest information to RPA pilots, commanders, and troops on the ground who, in turn, provide feedback on what they may be seeing. Thus, the RPA, as both a sensor and a shooter, is productive of a vast – and growing – telecommunications and surveillance infrastructure that underpins global relations of power (e.g., Gregory 2015; Grayson 2016).

Unlike other surveillance systems that can only provide short bursts of surveillance or periodic updates because of their positioning vis-à-vis their objects of study (e.g., satellites), the temporal limits of their operational range (e.g., spy planes), or difficulties with regards to concealment (e.g., human operators), the RPA, as shown in the previous chapter, is noted for its 'persistence'. Persistence within this discourse refers to its common meaning that denotes drive, tenacity, and single-minded will. This is different to the term's meaning within computing or technology discourses where persistence is used to denote a continuance of some factor after its cause has ceased to operate. While persistence is attributed to specific capabilities designed into RPAs, including the much vaunted 24 hrs plus of flight time, one could argue that the persistence of the systems is also a product of the way in which they are deployed. Long initial shifts for operators supplemented by operator hand-overs in mid-flight combine with specific rotations of air platforms so that one platform at the end of its shift can be replaced by another more fully fuelled system seamlessly in cases where a person or place of interest is being closely monitored (Williams 2011; Martin and Sasser 2010). Thus a quality that emerges from the complex interplay of machine–human interactions has become primarily located with the RPA itself.

Similarly, the weaponisation of the RPA has contributed to persistence being a primary frame of reference in associated techno-strategic discourses. The RPA is understood as being a unique node within the global communications network because of its 'shooter' capability. This means it can track/ monitor persons or places of interest in real time and complete the final act of the kill chain when a strike is deemed to be prudential. Even the use of the 'double-tap tactic' which fires a second round of missiles after surveying a strike site to ensure that a target has been eliminated can be said to contribute to representations of the RPA as persistent.[3] Tactics themselves become subsumed within the ontological framing of the RPA.

At the same time, RPAs can also be understood by operators and ground troops as a protective platform. Its role is to 'see over the next hill' and/or to identify what may be lurking around the corner so that ground troops are shielded from threatening surprises. In other words, the RPA functions to create a zone of reduced risk around ground troops positioned within what would otherwise be an uncertain battle-space. Thus, from this perspective, the RPA is less 'hunter-killer' and more 'guardian-nurturer', a means of protecting colleagues who might otherwise be vulnerable. When the RPA is engaged in hostilities, the rationale is that these actions are saving the lives of coalition troops, rather than avenging prior attacks or aggressively seeking to eliminate potential threats.

As suggested by Cohn (1987) in her analysis of nuclear weapons and the discourse of deterrence in the 1980s, gender is central to the operationalisation of techno-strategic discourse. With the development of remotely operated technologies, Mary Manjikian (2014: 51) has argued that it is important to determine whether these systems '…strengthen traditional conceptions of gender by creating a hypermasculine "super soldier" or [if] they undermine distinctions between the sexes as they create a fuzzy new set of genders and gender relations'. Particular characteristics of RPA systems are represented in such a way that qualities traditionally associated with masculinity are positioned as positive. In one respect, the RPA and the core characterisation of its persistence is no different in this regard. Moreover, as Benjamin (2012) has noted, the discursive representation of the RPA as a well-armed hunter-killer oozes with sexual imagery. Promotional materials from Lockheed-Martin, the manufacturer of the Hellfire missiles with which Predators are armed, equate the weaponisation of the RPA with sexual dynamism and the ability to engage in myriad forms of coitus:

> As in previous HELLFIRE II models, the HELLFIRE Romeo can lock onto targets before or after launch for increased platform survivability. But its new three-axis inertial measurement unit also enables properly equipped launch platforms to engage targets to the side and behind them without manoeuvring into position. It can be launched from higher altitudes due to its enhanced guidance system and improved navigation capabilities, which increases the missile's impact angle and enhancing lethality. A new multi-purpose warhead enables the HELLFIRE Romeo to defeat hard, soft and enclosed targets, thus allowing pilots to meet many contingencies with a single HELLFIRE loadout – it's one missile for many missions.
>
> (Lockheed Martin quoted in Benjamin 2012: location 607)

But the notion of the RPA as a vigilant guardian would also seem to draw from ideas traditionally associated with femininity, specifically motherhood. In this sense, the RPA fosters life rather than merely taking it away. Thus, the positive subjectivities that can be assumed by the RPA involve potentially contradictory gender representations. And while these stem in part from the potential of the RPA as a technology, it is also shaped by broader cultural understandings of gender subjectivities (Manjikian 2014: 49). Rather than being a liability, the potential ambiguously gendered subjectivity of the RPA is an asset, allowing for specific gendered representations to be drawn upon depending on the context and audience. For security hawks, the persistent hunter-killer subjectivity provides a level of gravitas and cache for the platform. For those who might be less inclined towards supporting – or deploying – systems that appear primarily oriented towards aggressive or hostile operations, the RPA as a watcher-guardian presents a more acceptable face.

Negative representations of RPA subjectivity in techno-strategic discourses are also gendered. Primarily, these stress the temperamental and vulnerable aspects of RPA platforms, characteristics traditionally associated with femininity – including its telecommunication streams – rather than forms of hunter-killer resiliency associated with masculinity. Strikingly, many of these negative characteristics stem from what are otherwise presented as strengths of these systems. First, are the actual difficulties in the command and control of RPA platforms. The ability to pilot these aircraft from long distances comes with a host of challenges. It is reported that communication disconnections are relatively frequent, including during decisive operational moments.[4] Even at the best of times, control through the Ku-band leads to a consistent delay of up to two seconds between a manoeuvre being initiated by the pilot and performed by the RPA. Both the time lag and periodic loss of command and control when communications go offline have necessitated that all take-off and landings be performed on-site using a different radio communications system.[5]

There are additional communications vulnerabilities that are associated with the RPA. Although command and control systems are supposed to be encrypted to prevent hostile take-over or sabotage, the data streams that come from drones until recently were not. In 2009, US forces discovered that Iraqi insurgents, using software that could be purchased for $26, had captured hours of video coverage (Gorman et al. 2009). More recently, even the claims about the security of command and control were called into question. In June 2012, the Department of Defence was humiliated after making a wager that a group of research scientists would not be able to remotely disrupt a RPA. They were able to do so by hacking into the RPA's GPS sensor and sending fake signals to confuse the navigation system (Mixon 2012). It is thought that an American RPA crash that occurred earlier that year in Iran may have been initiated by a similar form of attack. Both incidents have led to calls for RPAs to be outfitted with more robust GPS encryption and jamming technology. But encryption can only mitigate so many vulnerabilities that arise from RPAs being a node in a global communications network. For example, in September 2011, it was discovered that a logging virus of unknown origin had infected piloting systems at Creech Airforce Base Nevada. This meant that every keystroke in 'chat room' communications between RPA operators, troops on the ground, intelligence officers, air traffic controllers, and even commanders had been recorded by an unknown third party (Shachtman 2011).

In addition to communications vulnerabilities that make RPAs difficult to control, inherent design features including long glider-like wings, a lack of de-icing capabilities, and poor pilot visibility through the nose camera (even onsite pilots fly via camera) contribute to understandings of RPAs (and particularly the MQ-1 Predator) as a temperamental system. These elements have combined with the structural fragility of the airframe itself – which is a by-product of using light-weight materials to reduce fuel consumption and increase the range – to produce a platform that is more accident prone and less resilient than other aircraft deployed by the military. For example, in 2009, the USAF admitted that approximately one-third of their MQ-1 fleet had crashed (Drew 2009 quoted in Benjamin 2012: loc 324). In 2012, A BGOV report provided data that showed RPAs were the most accident prone systems in the USAF fleet. Global Hawks, Predators, and Reapers had a combined accident rate of 9.31 incidents per 100,000 hours of flight. Not only was this the highest rate for any system used in active duty but incidents were occurring at triple the USAF mean. Global Hawks, Predators, and Reapers had been involved in 129 accidents that resulted in at least $500,000 worth of damage and/or the loss of an aircraft over 15 years of combined service. While RPA supporters argued that accident rates for these platforms experienced significant reductions over time (e.g., in the fiscal year 2001, the accident rate for RPAs was 62.06 in combat operations which had dropped to 5.13 by 2011), critics still pointed out that these airframes experienced higher mishaps rates than piloted systems like the F-16 at similar points in their service history (McGarry 2012).

Much of what has been mentioned above are those capabilities and limitations that are not entirely unexpected. But as Jane Bennett (2005) has suggested in her analysis of the electrical blackout that encompassed much of the North American eastern seaboard in 2003, it might be that the actant subjectivity of the drone is initiated at those moments when it exceeds the parameters of its designed and/or programmed capabilities. The implication is that technology's ability to supersede the limits imposed upon it (i.e., 'going rogue') becomes both a subject position (i.e., the rogue) and an exercise in asserting agency that confirms its subjectivity. With RPAs there have been three widely reported incidents where RPAs have unexpectedly exceeded their limits and broken free of human command and control that prove illustrative. The first took place in 2009 when a Reaper, engaged in operations, went rogue over the skies of Afghanistan with a full payload of weapons. Unable to re-establish control over the platform, the Reaper was shot down by a regular fighter aircraft before it could enter the sovereign airspace of a neighbouring state. No explanation was offered for what might have contributed to the incident (Hsu 2009). In 2010, during a training session, a Fire Scout robo-chopper went 'rogue' in the Washington DC area. Communications between operator and platform were initially interrupted. But rather than assuming a holding pattern as programmed to do when communications go offline, the unit overrode this directive and proceeded into restricted airspace. The actions were blamed on a 'software anomaly'. Ground operators were also implicated for launching a command just prior to the communications interruption. This command was argued to be a contributing factor in the drone abandoning its holding pattern though no definitive evidence for this assertion was offered (Page 2010).

Perhaps the incident demonstrating the most outward expression of agential qualities took place in March 2011, at Camp Lemonnier in Djibouti. While stationed on the ground, a Predator started its own engines. This was all the more remarkable given that its fuel line had been disconnected and its ignition was manually disengaged (Whitlock 2012). In the text contained within the declassified incident report a USAF squadron commander stated, 'after that whole starting-itself incident, we were fairly wary of the aircraft and watched it pretty closely...Right now, I still think the software is not good' (Whitlock 2012).

Thus, we see an interesting dynamic in action. On the one hand, RPAs are understood and presented as the epitome of technological sophistication necessary for modern net-centric warfare. They are positively imbued with particular qualities like sensor-shooter, persistence, and hunter-killer. They administer kinetic force that is precise and that can be controlled. On the other hand, the RPA is also understood as being unusually vulnerable with these vulnerabilities stemming from root strengths identified with its material capabilities. They are also imbued with negative qualities such as being temperamental that make the periodic loss of control an ongoing problem. As Cohn (1987) has shown, these understandings are gendered, a dynamic that can give added depth to their descriptive claims. These practices though are not unique to the ways in which technologies are positioned within techno-strategic discourses. However, they provide an initial indication of the complexity in terms of the disjuncture between representations of RPAs in popular narratives and their actual capabilities and performance to date. These tensions also demonstrate the gendered fault-lines that traverse the materiality of the RPA as well as the multiple identities that an aesthetic subject may assume within a given context. It also speaks to the sites and moments where the drone may directly impact upon the sensory experiences of the human being – as will be discussed below.

RPA operators as aesthetic subjects

In her extensive study on killing in modern warfare, Joanna Bourke (1999: 5) notes that traditionally two central processes have been considered necessary to enable violence: 'numbed consciences' and 'agentic modes'.[6] When combined with what Robert Jay Lifton and Eric Markusen (1990) referred to as the 'technological imperative' – i.e., the desire to make full use of technologies that enable killing – these processes were said to be core authors of the horrific violence of twentieth-century war-fighting. Soldiers were argued to be socially detached, if not robotic, purveyors of death. Yet, Bourke was able to demonstrate that these assumptions did not necessarily hold when one looked more closely at how combatants understood killing. As her historical survey indicates, 'what is striking is the extent to which combatants insisted upon emotional relationships and responsibility, *despite* the distancing effect of much technology' (Bourke 1999: 7). Those who killed were often far from numb; they attempted to navigate the ethics and implications of their actions in combat by embracing killing. One of the means to this ends was to personalise the enemy as a way of forming 'a buffer against numbing brutality' (Bourke 1999: 7).[7] Another was to adopt what Bourke describes as the 'warrior persona', a persona that emphasised 'the chivalry, intimacy, and skill' required of combatants in order to kill (Bourke 1999: 51).

At the same time, Bourke's investigation reveals that forms of pleasure were often also experienced in the act of killing and in the preparations that precede it. Part of this pleasure derived from the intimacy of the act and the ways of seeing that enabled its commissioning. In her discussion of snipers, Bourke argues that 'the person doing the sniping…gained pleasure from his activities precisely because the foe was identifiable. Sniping too, was "personal" killing: it required that the killer looked his victim "in the eye"' (Bourke 1999: 61). Bourke demonstrates that the role of sight in orienting personal killing within a larger narrative and permitting forms of pleasure to be derived – whether through a sense of accomplishment, skill, or existentialism – remained an important element of modern warfare, even as technologies were developed that allowed combatants to be located at greater and greater distances away from the enemy. A type of scopic regime thus developed amongst combatants. Bourke (1999: 26) argues that during WWI, WWII, and Vietnam, combatants often '…interpreted their battleground experiences through the lens of an imaginary camera'. This not only positioned combatants as the chief protagonists in a personal (re)production of war, it also helped them to locate their own actions and feelings within codes that had been normalised through cinematic and photographic representations of combat. Thus, modes of killing and modes of seeing deployed by combatants in the wars of the twentieth century produced complex aesthetic subjects that could not be easily captured within the ideological framings used by proponents or opponents of militarism alike.

In this section, the pioneering work of Bourke is used as a foundation to analyse RPA operators as aesthetic subjects. Similar to Bourke's dismissal of orthodox understandings of combatants, I begin from the starting point that the popular characterisation of RPA pilots as detached killers devoid of emotional investment whose actions are guided by a 'playstation mentality' is not helpful. Instead, I explore the forms of affect and embodiment that are produced at the interface of the RPA operator and the RPA to show the power-relations and socio-cultural dynamics in this emerging site of modern warfare. This demonstrates that the materiality of the RPA and its systems are important.

Producing the optimal RPA pilot

The use of remotely piloted systems in combat is a relatively new phenomenon in military history. Just as strategies and tactics continue to be developed to maximise the effectiveness of RPAs, new recruitment procedures and training programmes continue to evolve in order to produce the most highly skilled RPA pilots and sensor operators possible. Through its training and certification programmes, the US military claims to be seeking to produce a particular kind of subject: caring, courteous, conscientious, and resilient. Pilots themselves while trying to live up to this ideal subjectivity find that there are competing affective experiences that are enabled through the RPA, its sensor technologies, and the ways in which these are deployed in battle-spaces. These produce forms of pleasure and anxiety that can be in excess of what is (or even can be) officially acknowledged. Moreover, these pleasures and anxieties are embodied. Their production is facilitated by interactions (and non-interactions) between the RPA, RPA operators, and other actors within the operational environment.

The production of RPA pilots and sensor operators is new ground for militaries around the world. Although the specifics of military training have changed over time, there has been less oscillation in terms of what qualities have been considered to make for a good soldier or good pilot. Over the course of the twentieth century, these qualities went beyond encompassing physical and intellectual requirements that were considered necessary to produce well-trained military personnel to include psychological and emotional profiles that were thought necessary to create a resilient fighting force (e.g., see Howell 2012). RPAs are interesting in that their material qualities and forms of deployment have meant that militaries are less certain about who makes for a good operator. Much of the contemporary public discussion (and criticism) has centred on whether it is necessary or desirable for a RPA operator to be a fully trained pilot in regular airframes and how much RPA specific training is required before an operator can be placed in combat situations (Atherton 2013; Copping 2013). While these are not unimportant issues, what has gone largely unnoticed is that militaries themselves are also engaged in research to determine what kind of subject is best suited to drone warfare. To these ends, the USAF has commissioned two recent studies that examined the psychological attributes important and/or critical to RPA pilots and sensor operators (McMillan et al. 2010; see also Chappelle et al. 2011).

As hopefully became clear from the discussion of the material characteristics of the RPA above, these platforms are extremely difficult to fly. Moreover, as a prominent sensor-shooter within a global information network, data streams, command and control directives, and events in the battle-space all come to impact upon pilots and operators at high, yet unsynchronised, tempos. Therefore, according to one study on RPA operators commissioned by the USAF:

> This position requires the pilot to visually discriminate and synthesize various images and complex data on several electronic screens while maintaining heightened vigilance to numerous sources of visual and auditory information necessary for maintaining situational and spatial awareness. The pilot must attend to visual-spatial two dimension input while performing numerical calculations for maneuvering [sic] the aircraft in addition to sustaining vigilance to multiple sources of visual and sensory input. The pilot must be attentive to several procedural checklists and processes with advanced computer systems while simultaneously translating

two dimensional information from video screens into spatial imagery. Despite the automated nature of many of the operations, the pilot in many situations must manually maneuver [sic] the aircraft (e.g., strategic deployment of weapons, BDA, positioning of surveillance, avoidance of bad weather, controlling the aircraft during equipment or systems failures, etc.). In short, pilots must rely upon a wide range of cognitive aptitudes when carrying out their duties in a confined environment with specific rules of engagement, tactics, and techniques.

(Chappelle et al. 2011: 5)

To function within such a pressurised environment, a series of cognitive abilities, psycho-motor skills, and personality traits have been identified as indicating potential suitability for RPA platforms. In terms of cognitive abilities, general cognitive ability, the speed and accuracy of information processing, visual perception – including acuity, scanning, discrimination, recognition, tracking, and analysis, spatial processing, such as the ability to construct 4-D mental representations from 2-D images, forms of memory, working, immediate, and delayed – across the senses, deductive reasoning, task prioritization, risk assessment, and cognitive flexibility were identified as critically important (Chappelle et al. 2011: 16). Key psychomotor processing attributes identified included fine motor dexterity, fast reaction time, and accurate psychomotor spatial coordination. Critical personality traits have been argued to include calculated risk taking, tolerance of stress, ability to work well with others in confined spaces, and social awareness (Chappelle et al. 2011: 16–18).

Counter-intuitively most research has concluded that psychomotor skills, while important, are not crucial to producing successful RPA pilots. In a qualitative study that relied on interviews with 82 officers in the USAF who either fly RPAs or command RPA squadrons, it was found that the cognitive and personality traits outlined above were considered to be critical elements with physical forms of dexterity and coordination less highly ranked. Moreover, there was a belief that as more and more RPA functions become automated, the importance of '...higher level information processing attributes would increase in comparison to psychomotor skills' (Chappelle et al. 2011: 18).

In addition, operators need to possess key intrapersonal and interpersonal skills that allow them to navigate the environment within RPA command and control centres. Mirroring many of the criteria said to define contemporary working environments outside of the military, informants have argued that intra-personal qualities such as composure, resilience, self-certainty, conscientiousness, perseverance, being success-oriented, and adaptable are important to surviving the social, emotional, and mental rigors of a RPA posting (Chappelle et al. 2011: 19). Key interpersonal attributes identified as being essential to forging the strong working relationships necessary for effective group dynamics included communication, humility, extraversion, a team orientation, and judgement (Chappelle et al. 2011: 22).

Two further personal qualities that encompass mental, emotional, and physiological elements were argued to be essential. Chappelle et al. (2011) reported that RPA operators found that the ability to compartmentalise was highly desirable because it allowed pilots to be able to 'deploy' at work and return to their domestic lives at home on a daily basis without the emotional impacts of work creating relationship difficulties in their domestic lives.[8] In addition, with RPA units subject to shift work and demanding operational tempos, the importance of cognitive stamina, that is the ability to concentrate and remain focused in the presence of significant mental and physical stresses, was also noted as important (Chappelle et al. 2011: 18–19).

What becomes clear then is that the desirable characteristics for a RPA pilot encompass a diverse set of attributes that go well beyond what one might anticipate. As Martin's memoir of his time as a RPA pilot makes clear, the role potentially requires – amongst other things – mastering the unique challenges of flying RPA platforms, the capabilities to effectively deal with the visual recognition of death, the frustrations of working within a hierarchical environment, constant shift rotations, maintaining healthy domestic relationships, routine boredom, and the sensory imposition of colleagues with personal hygiene issues in enclosed spaces (Martin and Sasser 2010).

But as Chappelle et al. (2011) have argued, given that pilots who demonstrated essential attributes could still under-perform, cognitive and personality attributes might only demonstrate capability. Therefore, they argued that motivation might be a better indicator of who will ultimately be successful as a RPA pilot (Chappelle et al. 2011: 11). Critical motivations were divided into moral and occupational types. While informants stressed that a RPA pilot needed a belief system that was compatible with deploying weapons, the prime motivation for their deployment was argued not to be focused on revenge, vengeance, or the thrill of the kill. Rather, the primary moral directive was framed in terms of saving the lives of US and coalition forces (Chappelle et al. 2011: 25). It was argued that this framing was best supported by the possession of a 'strong sense of duty as an officer and a war-fighter' and/or religious and spiritual beliefs (Chappelle et al. 2011: 25).

The results of these studies shows the inadequacy of the 'playstation mentality' thesis – the notion that what makes for an effective RPA operator – and therefore what kind of pilot militaries are trying to produce – is a sense of callous detachment from one's actions and their consequences in the battle-space. Numbed consciences and agentic modes channelled through a technological imperative are not what services like the USAF are looking to cultivate. Second, it shows that the US military is attempting to create a pilot subjectivity whose dispositions and reactions can be rendered calculable within a pre-established affective and emotional register. The subject is also very much a subject who should find self-motivation from an ethic of personal responsibility for the well-being of others. The emphasis is on finding subjects with the motivations, resiliency, and the sense of purpose that can be further cultivated to produce effective RPA pilots. This requires that various forms of physical and psychological screening are undertaken to select those most suitable who are then enrolled in specific platform training programmes depending on their general level of flying experience.

Thus, this research provides fascinating insight into how US military services are conceptualising the production of RPA pilots and sensor operators. While the importance of platform specific training is noted, the current emphasis is on finding the right kinds of people to assume the roles. This is certainly not unique. Finding suitable candidates for positions is a concern for any organisation. But what is interesting in this particular case is the emphasis on seeking out a particular subject profile that can cope with the contingencies of chaoplexic war-fighting. In other words, the USAF – and potentially other services – realise that despite attempts to do so, they are unable to control the environment in which RPA pilots are operating and thus must find and develop subjects whose essential attributes lend themselves to reduced uncertainty over potential reactions to stimuli. The desire is to develop RPA pilots who ideally become the disembodied and calculable *homo oeconomicus* for whom killing – when it takes place – is to be understood as a means of saving lives.

Beyond homo oeconomicus

While the goal may be to create pilots whose reactions can be subject to calculation and control – particularly as proposals for new systems require platforms to take account of the affective experiences of pilots – there are affects and feelings not captured in the optimal pilot narrative outlined above. Anxieties result from the experience of disembodiment in the RPA piloting process and pleasures generated by the specific human-technological sensory interaction in the battle-space that are in excess of what can be directly calculated and controlled.

The physical demands of many types of combat and platform piloting are often thought to be challenging because of the ways in which they are embodied. For example, the inertia and acceleration that impact upon a fighter plane in flight also have a physiological affect on the pilot. The experiences of wind, turbulence, gravitation force, orientation, visual cues, and even the feel of particular mechanisms within the aircraft are also physically experienced by the pilot. With RPAs, pilots have fewer embodied linkages to the aircraft (McCurley and Maurer 2015: 26). On the one hand, this has been viewed as a benefit:

> As a pilot of the future *now*, I didn't have to worry about the physiological stresses of high-speed flight in my grounded zero-knot, one-G cockpit. I could focus on the tactical situation while backed up by a team of experts spanning the globe.
>
> (Martin and Sasser 2010: 310)

On the other hand, although '...the best Predator pilots *felt* the airplane, even though they weren't actually in it', the lack of specific forms of embodiment experienced in comparison to other air platforms poses challenges (Martin and Sasser 2010: 23). Martin has noted the loss of 'feel' built into the Ground Control system for the MQ-1 Predator. The control system is designed so that maintaining altitude and direction requires that the control stick and rudders be held in place. If the control stick is let lose, it is spring loaded so that it returns to a centre position, ensuring that the RPA quickly resumes level flying. As mentioned above, with all control instructions being processed through two separate computer systems (one in the GCS and one in the RPA), any pilot initiated command necessarily involves a lag before it is received and completed by the platform, a lag that increases when using Ku-band communications. Moreover, Martin and Sasser (2010) note the Predator itself flew more like a glider than a jet aircraft with its long wings – bereft of de-icing technologies – making it vulnerable to turbulence, strong winds, and low temperatures (see also McCurley and Maurer 2015: 251).

Although RPAs are renowned for their advanced visual sensor technologies, these have not been designed to help with the piloting experience. The nose camera is fixed making it impossible for a pilot to look elsewhere for orientation cues and the presence of dangers. The camera itself also only has a 30-degree field of view in comparison to the human eye which has a 50-degree field of view. Due to these limitations, Martin has described the experience of the Predator as being akin to 'trying to fly while looking through a soda straw. Like riding a roller coaster without being able to turn your head up or down' (Martin and Sasser 2010: 24). The act of piloting the RPA and maintaining control of it raises demands that cannot be consistently conquered by dexterity, coordination, or psychomotor skills. Thus, RPA piloting is a source of stress and anxiety, irrespective of the environments in which these platforms are operating. The comparable lack of tangible physicality in relation to regular air-platforms, the lack of key embodied stimuli, an awareness of the high accident rate and the knowledge of the costs of a mistake – whether it be financial, professional, collateral damage, or to colleagues through the absence of a RPA in a rota – produce considerable pressures. These pressures are compounded by the working conditions for RPA pilots. A recent USAF study has shown that nearly half of drone operators suffer from high operational stress, resulting from the physical pressures of long hours and frequent shift changes, which have continued, in part, because of chronic staffing shortages (Bumiller 2011; Dao 2013).

Additional anxieties have been shown to arise from the geographical compression that takes place when one is stationed in the United States but flying RPAs in other parts of the world. During his time a Predator pilot during the height of the Iraq and Afghanistan conflicts, Martin noted the disconnect between being involved in combat operations and having to fulfil normal domestic duties – like picking up a gallon of milk – on the way home afterwards (Martin and Sasser 2010: 2). While some media reports have presented the exposure to disturbing images from the battle-space – RPA operators often must confirm the results of various strikes – as a source of alienation from the civilian world outside of the airbase, this has not been confirmed by psychological studies undertaken on RPA pilots to date. Rather, conflicting feelings have been noted by RPA pilots about the vast distance between themselves and the existential risks of the theatre of operations. The idea that being a proper warrior entails being in close proximity to the action is a strong social norm within military services that is often internalised by those who operate RPA platforms. One operator has explained in an online forum that RPA crews are kept out of harm's way because, 'we are considered too valuable because our training takes too long to risk us being hurt/killed. Sounds kinda pansy-like, but its [sic] the unfortunate truth.' [9]

There are even feelings of guilt by some pilots that they can enjoy the simple pleasures of civilian life after a shift in comparison to colleagues on the front line. As one RPA pilot has stated:

> You feel bad. You don't feel worthy. I'm sitting there safe and sound, and those guys down there are in the thick of it, and I can have more impact than they can. It's almost like I don't feel like I deserve to be safe.
>
> (Bowden 2013)

That there may be something less heroic about being a RPA operator has been at the heart of recent inter-service debates over whether drone pilots should be eligible for service medals.[10]

At the same time, the efforts of RPA operators are appreciated by colleagues who are positioned in harm's way on the ground. As one soldier has remarked in an online chat with a drone pilot:

> But even though I've heard more than a few jokes about UAV [unmanned aerial vehicle] operators, we always enjoyed having you around. I never heard one complaint when we had someone in the sky watching over us. There's nothing quite like hearing that familiar buzz in the sky, or watching an FMV [full motion video] feed, to feel just a little bit more assured.[11]

As such, the gendered subject positions of RPA operators and inter-subjective understandings associated within them are highly complicated and contingent, producing additional anxieties about what it means to pilot these systems and what one becomes when doing so.

More surprising than the inability to capture anxieties experienced by RPA operators is that official understandings of pilot subjectivity neglect to explore the forms of pleasure produced through human–RPA interaction. Bourke's historical investigation of intimate killing demonstrates that killing from the air was often presented by participants as sublime (Bourke 1999: 58). From Bourke's examination, it becomes clear that aerial combat has always been imbued with its own unique experiences for combatants. In particular, it has been forms of vision enabled by aircraft in combat, particularly RPA platforms that have been designed primarily as a means of providing an enhanced visual overview of the battle-space, that have been a source of pleasure. In turn, the enhancement of vision facilitated by RPAs has produced its own power-relations in contemporary counter-insurgency.

While the centrality of vision to practices of power has been identified by the likes of Michel Foucault (1977; 1990) and Edward Said (1978), it is Allen Feldman (1997: 30) who suggests that, 'compulsory visibility is the rationality of state counterinsurgency'. The initiation and completion of the 'kill chain' is therefore predicated on forms of '...visual surveillance [that authorise] this form of bodily intervention' (Feldman 1997: 27). Targets must be visually identified, confirmed, tracked, and positioned relative to other people and material objects in the complex spaces of everyday socio-economic activity. Claims to the precise and discriminatory nature of RPA strikes rest on a consensus that the images produced through forms of technologically enhanced vision are accurate.[12] This consensus, in turn, is reliant on the operationalisation of a specific scopic regime. According to Feldman (1997: 30), a scopic regime refers to the means and techniques:

> ...that prescribe modes of seeing and object visibility and that proscribe or render untenable other modes and objects of perception. A scopic regime is an ensemble of practices and discourses that establish truth claims, typicality, and credibility of visual acts and objects and politically correct modes of seeing.

The scopic regime that operationalises the kill chain is one that is shaped by epistemological and aesthetic realism. The practice of seeing via the imagining technologies found on RPAs is verisimilitude. This scopic regime operates under the assumption that vision – through technological enhancement – can become an infallible sense that captures *the* physical world independently of any subjective perceptions that we may have of it. The desire is for ever more improvements in the type of images, their resolution, and the duration of time for which they capture a field. The disembodied, a-historical, and objective observation that helps to define this vision is symbolic of what Donna Haraway (1988: 581) refers to as the 'god trick of seeing everything from nowhere.' Yet, the 'god trick' perspective produces conflicting feelings in RPA pilots as the sublime and uncanny interact to produce an unprecedented view of the battle-space in near real time. Reflecting on his own experiences, Martin notes that:

> The suddenness of action played out long distance on computer screens left me feeling a bit stunned. A surreal experience. Almost like playing the computer game *Civilization*, in which you direct units and armies into battle. Except with real consequences. I felt electrified, adrenalized...It would take some time for the reality of what happened to sink in, for 'real' to become *real*.
>
> (Martin and Sasser 2010: 31)

Later, Martin reveals that with the view from the RPA, '...I truly felt like an omnipotent god with a god's seat above it all' (Martin and Sasser 2010: 121). It is from this god-like perspective in which feelings of omniscience and omnipotence are produced that individuals are identified, tracked, targeted, and killed. As Chad Harris (2006: 119) notes, in opening up the battle-space through the intensive forms of observation enabled by new technologies, this space is concurrently locked down by these practices of high tech surveillance and the forms of violence they enable. Within the unfolding of the kill chain from this unworldly vantage point, targets are positioned as both 'indubitable recordings of what...[was]... simply there and as heroic feats of technoscientific production' (Haraway 1988: 582). The scopic regime thus encapsulates what has been identified as a predominantly masculine way of seeing that has developed through Western traditions in the sciences and humanities, a perspective from which a hidden observer can adjudicate upon the ordering of a particular space (Berger 2008; Haraway 1988; O Tuathail 1996). This is what Denis Cosgrove (2001) has referred to as the 'Apollonian gaze', a gaze that imposes order upon a diverse multitude from a single perspective that sees all that needs to be seen (see also Kaplan 2006; Harris 2006). Such is the desire for an ever present and unwavering gaze, the US military uses the word 'blinking' to describe those times when a space of interest is not being monitored by a drone (ISR Task Force 2013).

As alluded to earlier, the interface between the RPA and its human interlocutors constructs a cyborg assemblage of seeing that surveys, targets, and exterminates (Williams 2011; Masters 2005). Advanced camera technologies – including the ability to perceive human heat signatures – combine with a human operator, and other forms of surveillance/intelligence to which the operator has been made privy, to produce an image of the field of operation within which surveillance, and potentially strikes, will take place.[13] In the processes of taking two-dimensional data from the RPA and transposing it into four dimensions in order to position targets, RPA crews shift visual recognition into a mode of physical extermination. Martin outlines these thought processes at the moments leading up to the final act of the kill chain:

> I was concentrating entirely on the shot and its technical aspects. Right range, right speed, locked in. The man wasn't *really* a human being. He was so far away and only a high-tech image on a computer screen. The moral aspects of it – that I was about to assassinate a fellow human being from ambush – didn't factor in. Not at the moment. Not yet.
>
> (Martin and Sasser 2010: 43)

Thus, a set of perspectives – geopolitical, ideological, moral – must be assumed – and/or sublimated – by RPA operators in order to enable the completion of the kill chain. In this kill-chain, human and machine are one. As Feldman (1997: 38) notes, '[the RPA becomes] a prosthetic that extends ideology and visions of history into the depth of the human body, leaving the dead and depicted in its wake'.

But the argument that those who commit targeted killing possess a detached and disembodied vision that enables clinical precision when circumstances justify its deployment belies that vision is also an affective sensibility. To paraphrase John Berger's (2008) observation on the history of female portraiture in Western art, the body of the targeted, in the roles as an object of surveillance and as a physical conduit for the inscription of military power, serves to flatter the one who can view. This mode of 'seeing without being seen' is constitutive of an asymmetrical power-relation that produces pleasure for the viewer and can empower acts of violence (Feldman 1997: 40). Martin has argued that:

> Flying the Predator allowed me the extraordinary perspective of being not only a 'combatant', albeit from 7,500 miles away, but also an observer with a broad overview. I saw the war first hand day after day as it unfolded. It seemed I was always watching in real time, hovering above, sometimes swooping down to join it like the predator I was.
>
> (Martin and Sasser 2010: 77)

His memoirs contain several instances where it becomes very clear to him that subjects of interest have no idea that they are being watched. Their lack of awareness is viewed with mixtures of bemusement, incredulity, and an air of superiority (Martin and Sasser 2010: 110). Even as insurgents became aware of how their movements and positions were being exposed by RPAs, Martin relates how the measures taken to circumvent surveillance such as duck blinds were too crude to fool the sensors on the MQ-1. While realism as a mode of seeing can be deployed as a means of supporting '…truth claims about the rationality and efficacy of political violence', it also produces a form of pleasure that can be addictive for the one with the privilege of viewing (Feldman 1997: 41). Seeing others who cannot see you, finding out their secrets while your own remain hidden can be an enjoyable experience.[14] Moreover, with the RPA, one is not just a passive observer, but an actor capable of unleashing considerable violence upon – an often unsuspecting – target. As Martin observed, 'sometimes I felt like God hurling thunderbolts from afar' (Martin and Sasser 2010: 3). The forms of vision enabled by the RPA allow one to see in detail the scope of the damage inflicted. And the experience of viewing is reported by some to be sublime, a mixture of the pleasure of the extraordinary with the displeasure of awe as one approaches the verge of their own powerlessness to control the consequences of an act (of violence). After an incident in which he launched a missile attack on a vehicle containing a suspected militant, Martin reported that, 'nobody in the GCS uttered a word. We merely watched the video, mesmerized, both awed and horrified by the carnage and this guy's dying effort' (Martin and Sasser 2010: 239). Similarly, one 19-year old operator describes the first time he was required to apply kinetic force thusly:

> 'I was kind of freaked out,' the pilot said. 'My whole body was shaking. It was something that was completely different. The first time doing it, it feels bad almost. It's not easy to take another person's life. It's tough to think about. A lot of guys were congratulating me, telling me, 'You protected them; you did your job. That's what you are trained to do, supposed to do,' so that was good reinforcement. But it's still tough.
>
> (Bowden 2013)

But not everyone is similarly affected. Another operator has remarked, 'I have…watched quite a few assholes take a missile to the face and every time I love my job just a little more.'[15]

Despite the pleasure and power it may convey, this ability to see via the cyborg assemblage of drone and human, or to gain maximum benefit from the enhancement of vision enabled within this assemblage, remains a technological and psychological challenge. Moreover, the design of the platforms themselves has considerable impact over the ability to see. As previously mentioned, RPA design has sought to make the platforms as light as possible to maximise their range. However, their light weight makes them particularly vulnerable to wind and turbulence (Edwards 2007: 34). Instability harms the quality of the visuals obtained through sensory equipment. Thus, there continues to be research on how to better construct a sharp image from a blurred original by factoring in the motion of the RPA (Edwards 2007: 35).

The focus on technological fixes as a means of providing a more precise or accurate view of the battle-space transposes older arguments about overcoming the fog of war into a new context. Both assume that there is some position/context in which vision can be unlimited and untainted. Moreover, it elides the fact that sensory output produced by RPAs must be interpreted by subjects who themselves are shaped by the operational context within which they are situated. Yet, the obsession is with improving the range of sensors and the amount of visual data that they collect. This obsession has impacts for the subjects who are tasked with deciphering their meaning. It has been estimated that the USAF alone processes almost 1,500 hours of video a day and 1,500 still images. By 2010, processing 'required about nineteen analysts per drone' (Bumiller and Shanker 2011 quoted in Benjamin 2012: location 306). But with new technologies like the 'Gorgon Stare' and ARGUS, apparatuses of cameras purportedly designed to capture the entire field of vision over large urban centres, the need for intelligence analysts to interpret the data is set to increase. Recent estimates are that 2,000 analysts will be needed per RPA with these forms of visualising technology (Benjamin 2012: location 309). Paradoxically, the volume of data undermines what has been presented as a primary advantage of the RPA: the speed with which it can gather real-time information from the battle-space and then act upon it. RPA operators already feel frustrated by imagery analysts at the intelligence hubs because the lag between providing visual data to them and when they are willing to confirm sensor input (Martin and Sasser 2010: 130).

In addition to the challenges posed by the volume of data being collected and the impacts on RPA operators, the mission types and forms of surveillance undertaken shape the outlooks of RPA pilots and sensor operators. There are considerable difficulties in identifying/distinguishing specific insurgents as well as insurgents from the general population. For Gregory (2011a: 191), it is these inherent limitations in vision produced by a platform that aspires to provide an uncompromised view of the battle-space that is one of the major differences between the type of combat undertaken by drones and the simulacra of combat in video games. Unlike combat, enemies are clearly marked in video games – either through colour coded indicators or Orientalist depictions.

The difficulties in identifying specific persons of interest in relation to the general population have led to operators being trained to look for behaviours that break with regular routines that have been previously observed or that fit into the insurgent 'pattern of life'.[16] To gain a sense of what might constitute everyday rhythms in a specific locale and the range of expected behaviours can be a long and tedious surveillance process. As Martin remarked, 'it was almost like watching some reality TV program that went on endlessly' (Martin and Sasser 2010: 78). Another pilot has stated that 'we see them playing with their dogs or doing their laundry. We know their patterns like our neighbour's patterns. We even go to their funerals' (Abé 2012 quoted in Gregory forthcoming). Yet in the end, this knowledge can only go so far and determining what behaviour might be suspicious is claimed to be primarily instinctive (Martin and Sasser 2010: 46). Martin describes this role as '...like a cop who had a sixth sense that somebody or something was out of place' (Martin and Sasser 2010: 81). But the hermeneutics of suspicion that is supposed to guide how RPA operators interpret the environments over which they patrol also feeds into feelings of paranoia. Thus, the circumference of the circle of suspicion grows ever wider, with everyone and everything getting positioned as a potential threat (Martin and Sasser 2010: 40). And as the circle of suspicion enlarges, so to does the probability of a mis-identification being made with fatal consequences. Therefore, while producing power-dynamics based on an asymmetries between seeing and being seen, the 'view from nowhere', can produce additional anxieties for the RPA operator. One pilot identified as Major Dan has explained:

> There is a very visceral connection to operations on the ground.... When you see combat, when you hear the guy you are supporting who is under fire, you hear the stress in his voice, you hear the emotions being passed over the radio, you see the tracers and rounds being fired, and when you are called upon to either fire a missile or drop a bomb, you witness the effects of that firepower.
>
> (Bowden 2013)

Therefore, Gregory (2011a: 197) suggests, being 'drawn into and captured by the visual field itself' can combine with feelings of affinity with ground troops and sense of responsibility for them to produce a proximity in which RPA strikes are initiated based on interpretations of operational spaces and their inhabitants that over-emphasise the potential risks of the operational environment. The result of these actions can be the extermination of civilians.

That drone warfare is embodied, visual, and pregnant with calculations of risk should not be surprising. It is literally war experienced through a camera and on a screen. The impacts of the variegated ways of seeing, the forms of embeddedness they produce, and the demands they generate is evident in both the continuation of civilian deaths from drone strikes and the rates of post-traumatic stress disorder amongst drone operators – similar to other combat pilots – that result from operational pressures (e.g., Dao 2013; Chappelle et al. 2014). Thus, the use of kinetic force does not require the dehumanisation of the enemy or a disregard for the consequences of drone strikes. What is necessary, however, is faith in visual omniscience, that at the pivotal moment, one can clearly see all that needs to be seen. Challenges to this way of seeing, what gets missed by this way of seeing – including the deprioritisation of sound outside of military radio communications – and how we are socialised into believing that visual omniscience is possible, are important sites of culturally infused politics.[17]

Populations of interest as aesthetic subjects

As shown above, the exercise of scopic power enabled by the interface of RPA technology and the human creates both pleasures and anxieties for RPA operators. Enhanced visual capabilities are tempered by deficiencies in how these capabilities can be mobilised and a loss of other embodied senses that comes from remote piloting. Yet, underpinning the scopic power of the RPA is a belief that the images it is able to produce in real time are themselves – despite any limitations – real and accurate configurations of what exists in the battle-space. Thus, the scopic power produced by the RPA is harnessed for the purpose of subjecting specific individuals, groups, and communities to sovereign forms of power that literally render final decisions over life and death.

From a Foucauldian analytic, the formulation of scopic power and its affects reveals the complex interplay of sovereign, disciplinary, and biopolitical forms of governance within the preferred modes of contemporary counter-insurgency where the goal is to shape environments so that insurgency can be pre-empted from 'becoming' (Anderson 2011: 233). Those who are deemed to be dangerous – from the construction of profiles and signatures that arise through various forms of knowledge – are to be made visible, tracked, and when conditions are favourable, eliminated in a spectacularly violent demon-stration of scopic power. Beyond the elimination of those who are deemed to be dangerous, it is argued that targeted killing attempts to create anxiety for potential targets (and wider populations) about being monitored so that pre-ferred norms are internalised and behaviour is shaped into forms more amenable to the pacification goals of the counter-insurgency project (Feldman 1997: 41). In this fashion, it joins other types of effects based operations like shock and awe that attempt to disrupt networks and circulatory flows by utilising awesome displays of military power to disturb regular modes of being.[18]

But, this view would seem to concede far too much to claims about the efficacy of RPA strikes for the tasks of counter-insurgency. Moreover, it potentially gives credence to the notion that reactions to RPAs and strikes can be controlled by counter-insurgents, ensuring that the right messages are received and mitigating the chances of misinterpretation or fallout. As Martin (Martin and Sasser 2010: 246) noted of those he observed via the optical sensors on the MQ-1:

> It was almost like they were ants down there, predictable in their behaviour to some degree of mathematical probability, no more aware of the Predator's presence than they were of the Almighty watching them.

By paying closer attention to how populations subject to these strikes negotiate the relations of power underpinning them, complex subject positions that extend beyond a victimised *homo oeconomicus* become evident. While anxiety is conspicuously present, initial research indicates that the means of managing the pressure often goes beyond acquiescing to the demands of counter-insurgency forces. Moreover, from a different subject position, one can see how the RPA-human interface produces additional affects that are embodied. Unfortunately, studies of the impact of RPA strikes as assessed by those who are subject to them are still few in number.[19] Therefore, this analysis can only be preliminary in its observations.

Effects and affects under the gaze of the drone

While the RPA is a conduit for scopic power in contemporary counter-insurgency by enabling specific forms of vision (discussed above), for those subject to the cyborg gaze of the drone, the primary sense through which surveillance registers is not necessarily vision, but sound. For example, residents of the northern regions of Pakistan refer to RPA platforms as '...*bangana* – a form of the Pasthun word for "wasp", in reference to the ubiquitous buzzing sound of the drones' (CIVIC 2010: 59). This sound forms the basis of a poem called 'Ghazal' written and posted online by Sayyed Abdullah Nazami, allegedly a member of the Taliban (see Van Linschoten and Kuehn 2012: 205). Likewise, in Gaza, the RPA is colloquially referred to as '*zenana* – an Arabic word referring to a wife's relentless nagging that Gazans have adopted to describe the drone's oppressive noise and their feelings about it' (Cook 2013).

In parts of Pakistan, residents claim that the sound of drones can sometimes be heard 24 hours a day, with multiple drones converging on sites of particular interest, often located far away from areas of overt military engagement between insurgents and the Pakistani Army (CIVIC 2010: 59). Accounts that stress the stealth of the RPA, and its ability to observe while unobserved, miss that the platform may reveal itself through other senses. Thus, the RPA as an auditory presence has specific effects on those populations over which it roams.

For example, the constant presence of RPAs in Gaza – a presence that is sonic and visual – has combined with full knowledge of their destructive kinetic capabilities.[20] This confluence of embodied historical experience has led to an epidemic of post-traumatic stress disorder amongst children living in Gaza (Thabet and Thabet 2015). It has also influenced everyday activities as people attempt to ensure that they do not look suspicious while undertaking menial tasks (Wilson 2011 quoted in Benjamin 2012: location 986). How paths of movement and spatial positioning in landscapes are influenced by viewing the self primarily through the imagined gaze and perceived psychological motivations of the drone operator are elegantly described in Atef Abu Saif's (2015) diaries of the 2014 invasion of Gaza. Moreover, there is a growing awareness amongst populations subject to counter-insurgency interventions that targeting practices for RPA strikes are often based on circumstantial evidence that is collected through 'pattern of life analysis' or observations of 'tangential interactions' with militants (CIVIC 2010: 59). Thus, the sound of RPAs in the vicinity, the knowledge that one may be observed, and an awareness that particular activities could be interpreted as hostile or suspicious creates a very difficult terrain to be navigated.

Those affected by strikes in Pakistan have argued that they feel the pressure of being caught between two sets of forces (i.e., the Taliban and various counter-insurgency organisations) that offer no positive value to them. Interviews conducted by the Campaign for Innocent Victims in Conflict (CIVIC) with those directly impacted by RPA strikes revealed that they are thought to be 'generally accurate' (CIVIC 2010: 61). But this impression of general accuracy was tempered by criticisms of what has also occurred when strikes take place including civilian deaths and injuries. Informants believed that strikes were an ineffective strategy for reducing insurgent capability and argued that the practice threatened to drive those initially opposed to the insurgency, towards supporting the Taliban and other extremists. Thus, the general sentiment reported was of opposition to US RPA strikes and demands for the practice to be stopped (CIVIC 2010: 60).

These demands for strikes to end though were not necessarily a result of informants being sympathetic to the Taliban or other insurgency groups. CIVIC documented a general unease towards the Taliban and frustrations at how Taliban activities placed civilians in danger. Subjects were well aware that strikes are opportunistic events and that opportunities best present themselves when insurgents are 'readily visible', a condition more likely to occur when positioned in civilian population centres as opposed to the relative safety of hideouts in the hills and mountainous regions (CIVIC 2010: 61). Similarly, complaints also arose of residences being targeted after families were forced to allow insurgents into their homes, resulting in strikes long after militants had left the area. In a report commissioned by Amnesty International (2013: 32), informants revealed how residents felt caught between the fear of being targeted by a drone attack and the potential reprisals for refusing to host members of armed groups. From these stories, measures undertaken in order to mitigate risk paradoxically may lead to the production of other forms of vulnerability. Thus, the aesthetic consequences of the strikes are directly related to complex everyday negotiations that take place in the difficult political environments that can be found in places like Afghanistan, Pakistan, Somalia, Gaza, and Yemen. And these consequences extend beyond the immediate kinetic and emotional impacts of living through an attack. As one informant stated, 'we fear that the drones will strike us again...my aged parents are often in a state of fear. We are depressed, anxious, and constantly remembering our deceased family members...It often compels me to leave this place' (CIVIC 2010: 62).

Vivienne Jabri (2010: 11) has argued that 'contestation is nowhere more apparent than in acknowledging the place of injury in violent practices'. To deny that another has suffered at one's own hands is not just to deny the corporeality of war: it is to deny the very humanity of the person that one has injured. In the terminology of Elaine Scarry (1985), it constitutes a second process of unmaking. RPA strikes, in downplaying or denying civilian casualties, have been a contributing element to this dynamic in the global counter-insurgency campaign. While acknowledgement and compensation for victims has been offered on some occasions in some theatres of operation – as a part of the new 'hearts and minds' brand of counter-insurgency – this has not been a consistent practice. All those who were interviewed by CIVIC argued that they were owed compensation by the Pakistani and/or American governments. Yet none had received any official acknowledgement that strikes had occurred, let alone post-strike support and financial compensation for the loss of housing, businesses, income, injuries, and deaths.[21] And the logics underpinning RPA strikes actually inhibit the process of acknowledgement and restitution that can be important aspects of coming to terms with traumatic loss.

First, as part of an undeclared 'everywhere war', strikes often take place in the jurisdiction of states in which the military presence of the United States is not formally acknowledged by the acting government (e.g., Pakistan and Yemen) or in others (e.g., Somalia) where US military activity is not readily admitted to by American authorities. In Pakistan, government denial takes place even as US strikes are guided by information that may have been procured through the Pakistani intelligence services and their impact is verified by government authorities (CIVIC 2010: 63). Second is the paradox of precision strikes. Those interviewed in Pakistan have noted that improvements in RPA strike accuracy make it even more difficult to obtain compensation as the assumption is that civilian casualties are highly improbable (CIVIC 2010: 63). Moreover, those who have been hit by RPA strikes have argued that the general acceptance of the precision of RPA munitions – even amongst populations of interest – has made it increasingly difficult to clear one's name after one's residence has been targeted (CIVIC 2010: 61; CIVIC 2012: 21). The assumption has become that someone must have been engaged in something untoward. Yet, those who have experienced losses from a drone strike do not meekly stand aside. For example, Amnesty International (2013: 40–41) has documented that protests often take place after strikes in North Waziristan and that these protests occur at a higher rate than those directed in the aftermath of violence inflicted by the Pakistani military or the Taliban.

Beyond the subject positions as a target for forms of intervention in counter-insurgency, a victim of counter-insurgency, or one who exercises agency in order to gain recognition for losses suffered, there is also a proactive subject who utilises the structures that make RPA strikes possible for their own ends. CIVIC (2010) has reported allegations that locator chips – small electronic signalling devices designed to track high value targets – are being redeployed to settle local disputes and conflicts.[22] Similarly, a NSA programme called Geo-Cell that uses signals surveillance technology to locate suspected insurgents through their mobile phone or SIM card for the purposes of killing them, is also prone to on-the-ground modification. As Jeremy Scahill and Glen Greenwald (2014) have reported, insurgents are defending themselves against this form of surveillance by using multiple SIM cards and frequently exchanging SIM cards. It does not go beyond the realm of possibility to anticipate that insecure mobile phones may be redeployed, either through carelessness or with the intent to achieve local political objectives. These activities demonstrate the multiple forms of agency that are shaping the ways in which RPA strikes are experienced. They also illustrate how the broader components of countering counter-insurgency may be redeployed at sites of intervention in ways that evade capture, despite the imposition of forms of visibility by drone's scopic regime.

Conclusions

This chapter has outlined the complex aesthetic subjects that are produced through RPA strikes in the global counter-insurgency campaign. It has argued that the various – and sometimes contradictory – subject positions that are enabled through drone strikes can restructure relations amongst those who perpetrate acts of violence, those who experience these acts of violence, and the technologies that facilitate these acts of violence. Whereas orthodox accounts often privilege the forms of instrumental rationality and cognitive reasoning that attempt to control the impacts of political violence, the account offered in this chapter has stressed that intimate killing involves affective and embodied experiences that cannot be fully captured and disciplined. Moreover, it has demonstrated that claims about the centrality of clinical detachment and/or a 'playstation mentality' to the production of subjects capable of initiating targeted killing strikes misses the complex interplay of anxieties and pleasure that affect RPA operators.

RPA operators and target populations both play central roles in facilitating the scopic regime underpinning global surveillance. The global surveillance regime in turn will prove to be an enduring legacy of RPA platforms regardless of how long they continue to be used for the purposes of targeted killing; their scopic power-relations produce asymmetries between those who are seen and those who see, but are not seen. As the analysis suggests, it is important to remain aware of the relations of power that are contributing to this regime and the subject positions that constitute it. Primarily, as suggested by Gregory (2010; 2011b), the blend of human and machine, the use of sensors to heighten the depth of the visual field, and the presentation of images produced on two-dimensional screens contribute to a way of seeing that underpins contemporary counter-insurgency. And this particular way of seeing, like all ways of seeing, is cultural. Similarly, the preferred subjectivities that are said to emerge through drone warfare are also cultural artefacts. Moreover, the desire to exercise control over the affective and embodied experiences of drone warfare – including injury – and the ways through which this control is to be achieved are also cultural. Thus, this analysis foregrounds the inevitable space that exists between what is experienced in drone warfare and how this experience can be articulated, and/or understood. This too is a battle-space in which the politics of global counter-insurgency is being waged.

Notes

1 The spatial dimensions are explored in a subsequent chapter.
2 Important philosophical issues are being elided for the sake of presenting the broader argument. Primarily, a discussion of what constitutes agency (e.g., what separates free will from programming within a machine) and whether the status of a subject requires the ability to exercise agency (and, if so, what is the minimum threshold for this agency?) are not going to be forthcoming. Instead of arguing for what the RPA 'is', I am exploring how the RPA is represented as a subject based on its material capabilities. The phenomenological impacts of these representations and material capabilities are then presented, both in terms of the RPA as an aesthetic subject and with regards to those subjects with which it interfaces.
3 Critics argue that the use of the double tap tactic means that increasingly rescuers are getting caught in the second missile strike.
4 In his memoirs about his time as a Predator pilot, Maj. Martin (Martin and Sasser 2010) recalls one incident in which he lost communication with his aircraft in the midst of a missile strike. When communications go offline, RPAs are programmed to assume a holding pattern until pilot control is once again resumed.
5 This requirement speaks to the current drive by the United States to establish a series of RPA bases in the territories of client states in Eastern Africa, the Middle East, and Central Asia.
6 Agentic modes refer to instances in which individuals defer to the authority of others and undertake actions that they might ordinarily find abhorrent (Bourke 1999: 5).
7 More recent ethnographic work on Israeli snipers has supported this claim. See Bar and Ben-Ari (2005).
8 For a fictionalised account of these challenges through the perspective of a female RPA operator, see playwright George Brant's (2013) script for *Grounded*.
9 See www.reddit.com/r/IAmA/comments/rf7vh/iama_uav_operator_for_the_us_army_amaa. Accesses 15 October 2015.

10 It has been reported that the morale of RPA pilots is often boosted by receiving emails expressing thanks from those ground troops who they have helped. See Bumiller (2011).

11 See www.reddit.com/r/IAmA/comments/rf7vh/iama_uav_operator_for_the_us_army_amaa. Accessed 15 October 2015.

12 As Gregory (2015) has suggested though, how these systems of visual enhancement work in practice does not always meet expectations or desires.

13 Reports indicate that several intelligence officers are often present in the control room in order to interpret the images provided by RPA sensors. See Webb et al. (2010: 31–39).

14 Martin appears to take great pleasure in revealing instances in which he was able to witness forms of sexual activity via his video monitoring screens. That he focuses on behaviour that his audience will interpret as being deviant feeds into Orientalist understandings of Muslim sexuality.

15 See www.reddit.com/r/IAmA/comments/rf7vh/iama_uav_operator_for_the_us_army_amaa. Accessed 15 October 2015.

16 These are known as signature strikes and are differentiated from personality strikes – instances when a specific individual has been targeted in advance.

17 The fact that on-the-ground listening capacity is not given the same priority requires further reflection. As Nasser Hussain (2013) has noted, 'although the pilots can hear ground commands, there is no microphone equivalent to the micro-scopic gaze of the drone's camera'.

18 As Coward (2013: 103) notes, in this strategic deployment of airpower, 'the infra-structural targeting of strategic bombing...hybridises...with the effects based, pre-cision targeting of network centric warfare to create an imaginary of warfare that retains total war's identification of the entirety of social infrastructure as a target set while also aiming to decisively affect morale and capacity through the destruction of the nodes that will have the greatest effect on both. It is thus total in its target set, selective in its strategic targeting of key nodes.'

19 This relates to the relative difficulties of conducting field work in areas of Afghanistan, Pakistan, Iraq, Somalia, Yemen, and Palestine that have found themselves positioned as the epicentres for global counter-insurgency actions.

20 Eyal Weizman (2007: 240–244) has documented how targeted killings are operationalised in the Israeli occupied territories.

21 Insurgents have also not normally provided compensation to those who have been victimised by RPA strikes (CIVIC 2010: 62–63).

22 Locator chips may be placed in places of residence, motor vehicles, mobile phones, or other personal effects. These may be planted by informants who have been paid to do so.

Bibliography

Abé, N. (2012) 'Dreams in Infrared: The Woes of an American Drone Operator'. *Der Spiegel*. 14 December. Available from: http://www.spiegel.de/international/world/pain-continues-after-war-for-american-drone-pilot-a-872726.html. Accessed 15 October 2015.

Alston, P. (2010) *Addendum to the Report of the Special Rapporteur on Extrajudicial, Summary, or Arbitrary Executions: Study on Targeted Killings*. New York: Human Rights Council, United Nations.

Amnesty International (2013) *Will I Be Next? US Drone Strikes In Pakistan*. London, Amnesty International Publications. Available from: https://www.amnestyusa.org/sites/default/files/asa330132013en.pdf. Accessed 15 October 2015.

Anderson, B. (2011) 'Facing the Future Enemy: US Counterinsurgency Doctrine and the Pre-insurgent', *Theory, Culture, & Society*, 28(7–8), 216–240.

Aradau, C. (2010) 'Security That Matters: Critical Infrastructure and Objects of Protection', *Security Dialogue*, 41(5), 491–514.

Arnold, M. (2003) 'On the Phenomenology of Technology: the "Janus-Faces" of Mobile Phones', *Information and Organization*, 13(4), 231–256.

Atherton, K. D. (2013) 'No One Wants to be a Drone Pilot, US Airforce Discovers', *Popsci.com*. 21 August. Available from: http://www.popsci.com/technology/article/2013-08/air-force-drone-program-too-unmanned-its-own-good. Accessed 15 October 2015.

Bar, N. and Ben-Ari, E. (2005) 'Israeli Snipers in the Al-Aqsa Intifada: Killing, Humanity and Lived Experience', *Third World Quarterly*, 26(1), 133–152.

Barad, K. (2003) 'Posthumanist Performativity: Toward an Understanding of How Matter Comes to Matter', *Signs*, 28(3), 801–831.

Benjamin, M. (2012) *Drone Warfare: Killing by Remote Control*, New York, OR Books.

Bennett, J. (2004) 'The Force of Things: Toward an Ecology of Matter', *Political Theory*, 32(3), 347–372.

Bennett, J. (2005) 'The Agency of Assemblages and the North American Blackout', *Public Culture*, 17(3), 445–465.

Bergen, P. and Rothenberg, D. (eds) (2014) *Drone Wars: Transforming Conflict, Law, and Policy*, Cambridge, Cambridge University Press.

Berger, J. (2008) *Ways of Seeing*, London, Penguin.

Bourke, J. (1999) *An Intimate History of Killing: Face to Face Killing in Twentieth-Century Warfare*, London, Granta Books.

Bowden, M. (2013) 'The Killing Machines: How to Think About Drones'. *The Atlantic*. September. Available from: http://www.theatlantic.com/magazine/archive/2013/09/the-killing-machines-how-to-think-about-drones/309434. Accessed 15 October 2015.

Brant, G. (2013) *Grounded*, London, Oberon Books.

Bumiller, E. (2011) 'Air Force Drone Operators Report High Levels of Stress'. *New York Times*. 18 December. Available from: http://www.nytimes.com/2011/12/19/world/asia/air-force-drone-operators-show-high-levels-of-stress.html?_r=0. Accessed 15 October 2015.

Bumiller, E. and Shanker, T. (2011) 'War Evolves With Drones, Some as Tiny as Bugs'. *New York Times*. 19 June. Available from: http://www.nytimes.com/2011/06/20/world/20drones.html?_r=0. Accessed 15 October 2015.

Campaign for Innocent Civilians in Conflict [CIVIC] (2010) *Civilians in Armed Conflict: Civilian Harm and Conflict in Northwest Pakistan*. Available from: http://civilia nsinconflict.org/uploads/files/publications/civilian_harm_in_nw_pakistan_oct_2010.pdf. Accessed 15 October 2015.

Campaign for Innocent Civilians in Conflict [CIVIC] (2012) *The Civilian Impact of Drones: Unexamined Costs, Unanswered Questions*. Available from: http://civiliansin conflict.org/.../The_Civilian_Impact_of_Drones_w_cover.pdf. Accessed 15 October 2015.

Chappelle, W., McDonald, K. and McMillan, K. (2011) *Important and Critical Psychological Attributes of USAF MQ-1 Predator and MQ-9 Reaper Pilots According to Subject Matter Experts*. US Air Force Research Laboratory. Wright-Patterson Air Force Base.

Chappelle, W. L., McDonald, K. D., Prince, L., Goodman, T., Ray-Sannerud, B. N. and Thompson, W. (2014) 'Symptoms of Psychological Distress and Post-traumatic Stress Disorder in United States Air Force "Drone" Operators', *Military Medicine*, 179(8S), 63–70.

Cohn, C. (1987) 'Sex and Death in the Rational World of Defense Intellectuals', *Signs*, 12(4), 687–718.

Cole, C., Dobbing, M. and Hailwood, A. (2010) *Convenient Killing: Armed Drones and the'Playstation'Mentality*, Oxford, Fellowship of Reconciliation England.

Cook, J. (2013) 'Gaza: Life and Death under Israel's Drones'. *Al Jazeera*. 28 November. Available from: http://www.aljazeera.com/indepth/features/2013/11/gaza-life-death-un der-israel-drones-20131125124214350423.html. Accessed 15 October 2015.

Copping, J. (2013) 'The First Deskbound Drone Pilots Get Their RAF Wings'. *The Telegraph*. 2 April. Available from: http://www.telegraph.co.uk/news/uknews/defence/ 9967186/The-first-deskbound-drone-pilots-get-their-RAF-wings.html. Accessed 15 October 2015.

Cosgrove, D. (2001) *Apollo's Eye: A Cartographic Genealogy of the Earth in the Western Imagination*, Baltimore, MD, Johns Hopkins University Press.

Coward, M. (2006) 'Against Anthropocentrism: The Destruction of the Built Environment as a Distinct form of Political Violence', *Review of International Studies*, 32(3), 419–437.

Coward, M. (2009) 'Network-Centric Violence, Critical Infrastructure and the Urbanization of Security', *Security Dialogue*, 40(4–5), 399–418.

Coward, M. (2013) 'Networks, Nodes and De-Territorialised Battlespace: The Scopic Regime of Rapid Dominance', in M. W. Peter Adey, and Alison J. Williams (eds), *From Above: The Politics and Practice of the View from the Skies*, London, Hurst and Company, 95–117.

Dao, J. (2013) 'Drone Pilots Are Found to Get Stress Disorders Much as Those in Combat Do'. *New York Times*. 22 February. Available from: http://www.nytimes. com/2013/02/23/us/drone-pilots-found-to-get-stress-disorders-much-as-those-in-com bat-do.html. Accessed 15 October 2015.

Drew, C. (2009) 'Drones are Weapons of Choice in Fighting Qaeda'. *New York Times*. 16 March. Available from: http://www.nytimes.com/2009/03/17/business/17uav.html. Accessed 15 October 2015.

Editors of Scientific American (ed.) (2013) *The Changing Face of War*, New York, Scientific American.

Edwards, C. (2007) 'The Automatic Airforce', *Engineering and Technology*, 2(4), 32–35.

Feldman, A. (1991) *Formations of Violence: The Narrative of the Body and Political Terror in Northern Ireland*, Chicago, University of Chicago Press.

Feldman, A. (1997) 'Violence and Vision: The Prosthetics and Aesthetics of Terror', *Public Culture*, 10(1), 24–60.

Foucault, M. (1977) *Discipline and Punish: The Birth of the Prison*, New York, Pantheon Books.

Foucault, M. (1990) *The History of Sexuality: An Introduction*, New York, Random House.

Gorman, S., Dreazen, Yochi J. and Cole, August (2009) 'Insurgents Hack U.S. Drones'. *The Wall Street Journal*. 17 December. Available from: http://online.wsj. com/article/SB126102247889095011.html. Accessed 15 October 2015.

Gregory, D. (2010) 'War and Peace', *Transactions of the Institute British Geographers*, 35, 154–186.

Gregory, D. (2011a) 'The Everywhere War', *The Geographical Journal*, 177(3), 238–250.

Gregory, D. (2011b) 'From a View to a Kill: Drones and Late Modern War', *Theory, Culture, & Society*, 28(7–8), 188–215.

Gregory, D. (2015) 'Angry Eyes'. *Geographical Imaginations*. Available from: http://geographicalimaginations.com/2015/10/01/angry-eyes-1. Accessed 15 October 2015.

Gregory, D. (Forthcoming) *Moving Targets and Violent Cartographies*. University of British Columbia.

Grayson, K. (2016) 'Drones: Political Economy, Networks, Bodies, Agency', in M. B. Salter (ed.), *Making Things International 2: Catalysts and Reactions*, Minneapolis, University of Minnesota Press.

Haraway, D. (1988) 'Situated Knowledges: The Science Question in Feminism and the Privilege of Partial Perspective', *Feminist Studies*, 14(3), 575–599.

Harris, C. (2006) 'The Omniscient Eye: Satellite Imagery, "Battlespace Awareness" and the Structures of the Imperial Gaze', *Surveillance and Society*, 4(1/2), 101–122.

Holmqvist, C. (2013) 'Undoing War: War Ontologies and the Materiality of Drone Warfare', *Millennium*, 41(3), 535–552.

Howell, A. (2012) 'The Demise of PTSD: From Governing through Trauma to Governing Resilience', *Alternatives: Global, Local, Political*, 37(3), 214–226.

Hsu, J. (2009) 'Air Force Shoots Down Runaway Drone Over Afghanistan'. *PopSci*. 14 September. Available from: http://www.popsci.com/military-aviation-amp-space/article/2009-09/when-drones-go-wild-air-force-shoots-them-down. Accessed 15 October 2015.

Hussain, N. (2013) 'The Sound of Terror: The Phenomology of a Drone Stike'. *Boston Review*. 16 October. Available from: http://www.bostonreview.net/world/hussain-drone-phenomenology. Accessed 15 October 2015.

ISR Task Force (2013) 'ISR Support to Small Footprint – CT Operations in Somalia/Yemen'. *Pentagon*. Available from: https://theintercept.com/document/2015/10/14/small-footprint-operations-2-13. Accessed 15 October 2015.

Jabri, V. (2010) *War and the Transformation of Global Politics*, Basingstoke, Palgrave Macmillan.

Kaplan, C. (2006) 'Mobility and War: The Cosmic View of US "Air Power"', *Environment and Planning A*, 38(2), 395–407.

Latour, B. (2005) *Reassembling the Social: An Introduction to Actor-Network-Theory*, Oxford, Oxford University Press.

Lifton, R. J. and Markusen, E. (1990) *The Genocidal Mentality: Nazi Holocaust and Nuclear Threat*, New York, Basic Books.

Manjikian, M. (2014) 'Becoming Unmanned: The Gendering of Lethal Autonomous Warfare Technology', *International Feminist Journal of Politics*, 16(1), 48–65.

Martin, M. and Sasser, C. (2010) *Predator: The Remote-Control Air War over Iraq and Afghanistan: A Pilot's Story*, Minneapolis, Zenieth Books.

Masters, C. (2005) 'Bodies of Technology: Cyborg Soldiers and Militarized Masculinities', *International Feminist Journal of Politics*, 7(1), 112–132.

McCurley, M. and Maurer, K. (2015) *Hunter Killer: Inside America's Unmanned Air War*, New York: Dutton.

McGarry, B. (2012) 'Drones Most Accident-Prone U.S. Air Force Craft: BGOV Barometer'. *Bloomberg*. 18 June. Available from: http://www.bloomberg.com/news/2012-06-18/drones-most-accident-prone-u-s-air-force-craft-bgov-barometer.html. Accessed 15 October 2015.

McMillan, K. K., Chappelle, Wayne, King, Ray, and McDonald, Kent (2010) *Psychological Profile of MQ-1 Predator and MQ-9 Reaper Pilots*. USAF School of Aerospace Medicine, Neuropsychiatry Branch.

Mitchell, T. (2002) *Rule of Experts: Egypt, Techno-Politics, Modernity*, Berkeley, University of California Press.

Mixon, M. (2012) 'Todd Humphreys' Research Team Demonstrates First Successful GPS Spoofing of UAV'. The University of Texas at Austin. Available from: http://www.ae.utexas.edu/news/archive/2012/todd-humphreys-research-team-demonstrates-first-successful-gps-spoofing-of-uav. Accessed 15 October 2015.

O Tuathail, G. (1996) *Critical Geopolitics*, Minneapolis, University of Minnesota Press.

Page, L. (2010) 'Robot Kill-Chopper Goes Rogue Above Washington DC'. *The Register*. 26 August. Available from: http://www.theregister.co.uk/2010/08/26/fire_scout_washington_hiccup. Accessed 15 October 2015.

Peoples, C. (2009) 'Technology, Philosophy and International Relations', *Cambridge Review of International Affairs*, 22(4), 559–561.

Rancière, J. (2006) *The Politics of Aesthetics*, London, Contiuum Books.

Rancière, J. (2010) *Dissensus: On Politics and Aesthetics*, London, Continuum.

Said, E. (1978) *Orientalism*, New York, Pantheon.

Saif, A. A. (2015) *The Drone Eats With Me: Diaries from a City Under Fire*, Manchester, Comma Press.

Scahill, J. and Greenwald, G. (2014) 'The NSA's Secret Role in the U.S. Assassination Program'. *The Intercept*. 10 February. Available from: https://firstlook.org/theintercept/article/2014/02/10/the-nsas-secret-role. Accessed 15 October 2015.

Scarry, E. (1985) *The Body in Pain: The Making and Unmaking of the World*, Oxford, Oxford University Press.

Shachtman, N. (2011) 'Exclusive: Computer Virus Hits U.S. Drone Fleet'. *Wired*. 7 October. Available from: http://www.wired.com/dangerroom/2011/10/virus-hits-drone-fleet. Accessed 15 October 2015.

Shapiro, M. J. (2010) *The Time of the City: Politics, Philosophy and Genre*, London: Routledge.

Shouse, E. (2005) 'Feeling, Emotion, Affect', *M/c Journal*, 8(6). Available from: http://www.journal.media-culture.org.au/0512/03-shouse.php. Accessed 15 October 2015.

Thabet, A. M. and Thabet, S. S. (2015) 'Trauma, PTSD, Anxiety, and Resilience in Palestinian Children in the Gaza Strip', *British Journal of Education, Society & Behavioural Science*, 11(1), 1–13.

van Linschoten, A. S. and Kuehn, F. (eds) (2012) *The Poetry of the Taliban*, London, Hurst & Company.

Walters, W. (2014) 'Drone Strikes, Dingpolitk and Beyond: Furthering the Debate on Materiality and Security', *Security Dialogue*, 45(2), 101–118.

Webb, D., Wirbel, L. and Sulzman, B. (2010) 'From Space, No One Can Watch You Die', *Peace Review*, 22(1), 31–39.

Weizman, E. (2007) *Hollow Land: Israel's Architecture of Occupation*, London, Verso Books.

Whitlock, C. (2012) 'Remote U.S. Base at Core of Secret Operations'. *The Washington Post*. 25 October. Available from: http://articles.washingtonpost.com/2012-10-25/world/35499227_1_drone-wars-drone-operations-military-base. Accessed 15 October 2015.

Williams, A. J. (2011) 'Enabling Persistent Presence? Performing the Embodied Geopolitics of Unmanned Aerial Vehicle Assemblage', *Political Geography*, 30, 381–390.

Wilson, J. R. (2002) 'UAVs and the Human Factor', *Aerospace America*, 40(7), 53–57.

Wilson, S. (2011) 'In Gaza, Lives Shaped by Drones'. *The Washington Post*. 3 December. Available from https://www.washingtonpost.com/world/national-security/in-gaza-lives-shaped-by-drones/2011/11/30/gIQAjaP6OO_story.html. Accessed 15 October 2015.

6 The quotidian geopolitics of targeted killing strikes

Introduction

It is not surprising that the use of remotely piloted aircraft (RPAs) to initiate missile strikes and engage in the targeted killing of persons of interest has drawn the attention of researchers in security studies, geopolitics, and international law. As shown in previous chapters, the increase in the prominence of RPAs has raised a series of contentious questions about the practices of counter-insurgency, the legitimacy of targeted killings and pre-emptive warfare, as well as provoking debates about the role of signature and personality strikes within counter-insurgency campaigns. And while these questions are indeed important, they often take the contexts within which RPAs are operating and the mechanisms of governance shaping their use as givens. In contrast, other sets of inquiries have sought to explore how the use of RPAs becomes possible within what Derek Gregory (2011a) has disparagingly referred to as the 'everywhere war'. Moreover, within this latter body of work, there has been a concern with how drone warfare is productive of broader territorial, aesthetic, cultural, biopolitical, and security dynamics shaping the contemporary theory/practice of geopolitics.[1]

In this chapter, I argue that while existing critical work has made a valuable contribution to understanding how drone warfare becomes possible, there has been an inadvertent focus on the exceptional spaces, places, and actions that are constitutive of the 'everywhere war'. Thus, as a supplement to these critical interrogations of the everywhere war, it is important to also examine the quotidian geopolitical places that make the everywhere war an 'everyday war'. Violent practices that emerge from the targeted killing assemblage contribute to processes of territorialisation that define spatial relations at the level of the quotidian. In this sense, it is important to map how the expanding battle-space mobilised through the targeted killing assemblage colonises places and seeks to disrupt their temporalities, whether transcendental or the rhythmic time of the everyday. In invoking 'the everyday' and quotidian, I am not making an implicit normative claim about their authenticity or preferential status in relation to the international.[2] Rather, I am deploying them to identify a particular space that counter-insurgency seeks to render amenable to its strategies of governance while those who occupy this space negotiate various ways of being. Therefore, the quotidian is a location where the spatial abstractions of counter-insurgency collide with the absolute spaces of habitants, that is, social worlds filled with places invested with (contested) meaning(s) (Davies forthcoming).

To begin mapping the quotidian geopolitics of the RPA, this chapter starts with an examination of the 'vertical geopolitics' and security imaginaries that are contributing to conceptualisations of sovereignty and territoriality at the heart of the everywhere war. These spatial analyses draw our attention to the current geopolitical prominence being given to 'ungoverned spaces' and the ways in which actors like the United States are attempting to govern these spaces. The analysis will then move to consider what spatial relationships have not yet been captured by such accounts. Two related places that are representative of the quotidian geopolitics of the everywhere war are advanced. The first is sacred places and how these are being problematised within the counter-insurgency doctrines being used to administer and guide RPA strikes. It will be argued that overly rigid understandings informing counter-insurgency take a limited view of what constitutes a sacred place and ignore how the contested borders of private/public spatial distinctions may influence understandings of political violence and perceptions of legitimacy. The second is to turn to one of the everyday places at the core of the everywhere war and in particular with RPA deployments: the home. It will be suggested that core dynamics central to the production of the aerial dimensions of the everywhere war can be found in homes. This will be explored by unpacking how drone strikes have undermined the sacred qualities of the home in theatres of counter-insurgency. Katherine Brickell (2012: 566–67) argues that 'the balance of geographical and social science literature still gravitates towards the impact of disembodied state-centred geopolitics *on* home life (thereby privileging big "P" politics)'. While I suspect I will not address this to the extent intended by Brickell's provocation, I do hope to make the '…geopolitical imperative to rhetorically divorce or downplay the effects of modern warfare on domestic life' more difficult (Brickell 2012: 578). But before exploring how the targeted killing assemblage colonises sacred space, the RPA must be placed in a wider geographical imagination.

Vertical geopolitics, ungoverned spaces, and the everywhere war

Stephen Graham (2004: 20) has argued that traditional understandings of geopolitics assumed the presence of '…Euclidean territorial units jostling for space on contiguous maps'. Thus, surfaces were attributed with a flatness and smoothness that ignored the various elevations at which key elements of geopolitics, from bordering practices to war-fighting, take place. Moreover, for those interested in problematising contemporary security logics, two-dimensional imaginaries neither reflect how advanced militaries are theorising counter-insurgency warfare nor how it is practiced (and resisted) in the urban areas of the global south (e.g., Adey 2010; Graham 2011; US Armed Forces 2005a). Drawing from the theoretical insights of Delueze and Guattari as well as the pioneering work of Eyal Weizman on Israel–Palestine, Graham (2004: 20) argued that geopolitics should develop '...a three-dimensional view of space-time...' that would comprise a '(geo)politics of verticality'. Attention to the vertical plane would reveal the kinetic, (bio)political, and affective dimensions of aerospace power being used to acquire, manage, and control territories, resources, and populations located across the horizontal plane. Graham also argued that critical examinations of vertical geopolitics would help to disturb key binaries that constrict geopolitical discourses such as international and national, domestic and foreign, and military and civil by highlighting how vertically produced scalar dynamics and shared practices travel across all of the planes of spatial analysis. Finally Graham's call for an interrogation of vertical geopolitics was also a call for a greater sensitivity to how militarism and its territorial projects are produced, sustained, and operationalised in the twenty-first century.

Graham's work spurred a renewed interest within the field of geopolitics in the examination of how existing and emerging technologies were reconfiguring the organisation, management, and control of space along vertical axes that extended into the stratosphere to deep down below the surface of the ground, and even into the physiology of the human body.[3] Thus, this research highlighted how the vertical geopolitical gaze extended across horizontal and vertical planes as well as through time to produce relations of power and particular subject positions like insurgent, terrorist, or civilian.

While the contemporary importance of vertical geopolitics extends beyond the global counter-insurgency campaign, its practices have been shaped by counter-insurgency's attendant discursive formations. Narrow readings of (geo)political history have been mobilised as a tool to understand the fault-lines of contemporary conflict and to make analogies regarding existing distributions of power.[4] Concurrently, advanced actuarial, epidemiological, and environmental data have been analysed to discover the sources of future disruption for the purposes of pre-empting their outward manifestation.[5] Within this discursive formation, the primary emphasis of geopolitical logics, as practiced by wealthy industrialised states with expeditionary military capability, has shifted. As Katharyn Mitchell (2010: 289) has noted, global geopolitics as practiced by liberal powers has transformed from a concern with the *effective* containment of dangerous spaces, the prevalent logic during the Cold War, to the *effective* administration of dangerous spaces through direct and indirect forms of governance. Moreover, mirroring changes in criminological theory that have become prominent with the socio-economic dislocations engendered by the rise of neoliberalisation, she notes that the tolerance for behaviours perceived as deviant has greatly diminished; just as J.Q. Wilson argued that the presence of graffiti leads down a slippery slope to social anarchy, so too does the existence of particular ways of life in regions far away from one's territorial borders which are perceived as potentially imminent existential threats. Without a doubt, the events of September 11, 2001 reinforced this geographical imaginary.

The concern with ungoverned spaces and those who inhabit them has enabled advanced militaries to operationalise an expanded global threat matrix. The construction of threats, what they are said to threaten, and what must be protected from them, involve complex combinations of performativity and performances within the matrix to render a 'geography of evil' (Campbell 1998: 88). Gregory (2010: 172–173) has argued that such imaginaries and the political clout to act upon them relies on a series of spatial performances that define enemies as well as shape the means of war. His typology distinguishes amongst three central spatial practices: locating, inverting, and excepting. Locating takes place in the abstract space that emerges along the techno-cultural register. It is produced through practices such as cartography, threat assessment, and forms of targeting central to the everywhere war. Inverting is the process through which others are produced as radically alien though cultural-political processes. Alien others become known by harnessing an emerging regime of truth that draws upon a mixture of scientific, psychological, historical, and anthropological forms of knowledge conditioned by the problematique of control. These populations are then positioned within spaces that are understood as irredeemably violent in contradistinction to the 'open, unitary, and generous' space of the Self (see also Springer 2011). Decisions, based on risk assessments derived from the forms of power knowledge harnessed by the counter-insurgency machine are then used to initiate practices of conversion, quarantine, termination, or neglect as deemed necessary. Within these practices, Gregory (2010), via Agamben (1998), highlights the central role of the exception, the politico-juridical mechanisms that allow categories of people – and those who govern them – to be placed outside the rules of law, which permits measures that would ordinarily be verboten (see chapter 2 in this volume for a critique). Concurrently, novel forms of violence that arise from contemporary process of locating, inverting, and excepting – such as drone warfare – expand the scope of what is considered legally permissible (Morrissey 2011; Jones 2015).

But legal permissibility has blossomed into a new understanding of humanitarian legitimacy. The widening scope and hybrid forms of so-called new wars, humanitarian wars, resource wars, drug wars, urban wars, and counter-insurgency wars constitutive of the 'everywhere war' provide a pressing political imperative to understand how techno-cultural assemblages in general – and those specifically organised along a vertical axis – could contribute to modes of governance and control that have begun to be conceived on a planetary scale (Grayson 2012b: 124). As has been shown previously in chapter 4, styles of thought drawn from the realm of science such as chaoplexity, network analysis, and the swarm have been inserted into the longer-standing paradigm of total war (Bousquet 2009). Verticality, while always present in military thinking, now assumes an essential role in modern battle-spaces where technologically advanced forces attempt to impose an order on what have been represented as disorderly subjects (Neocleous 2013). The ability to govern from the air with surveillance capabilities and precision weapons systems that can target 'illegitimate' forms of life have perversely been invested with a humanitarian legitimacy not granted to other types of combat or armed coercion because of perceptions that these forms of war are less brutal and capricious as well as more proportionate and precise (Strawser 2013).

As has been argued above, the use of RPAs within the targeted killing assemblage to govern is imbricated within calculations of security, territory, and population (Foucault 2009). These in turn are produced – in part – through a geographical imagination that redeploys old elements of geopolitical thinking within a broader (re)configuration of a global battle-space. The RPA reignites concerns about strategic territorial acquisition for the purposes of projecting military power. For example, the United States has been securing land, leases, and agreements in states like Djibouti, once among the backwaters of American global geostrategy, in order to build airfields specifically for RPA fleets. The re-emergence of the horn of Africa to heightened geopolitical significance can be linked to the imperative to bring order to spaces perceived to be hot-beds of insurgency – e.g., Somalia and Yemen – as well as the current cruising ranges of RPAs themselves. Maps of American-controlled airbases in the Middle East and East Africa have been depicted as nodes in a series of concentric circles that overlap above current theatres of counter-insurgency operations. Similarly, the RPA has also initiated a new geopolitical resource calculus. While the role of hydro-carbon acquisition in US strategic thinking is well recognised, the drive to increase the production of RPAs – and their constitutive components – has catalysed a search for new sources of heavy rare earths required for the manufacture of propulsion and targeting systems (Mulvany 2012). With China currently representing both the largest supplier and consumer of rare earths, spatial understandings that enable RPA warfare have thus become entangled within the discursive formation of the 'China Threat'.

But alongside traditional elements of geopolitical understanding, RPA technology has practically problematised what John Agnew (1994) once referred to as the 'territorial trap'. RPAs undermine any lingering notion that states might be containers and instead demonstrate that multiple network links (e.g., political, economic, and infrastructural) with varying scalar connections are required to conduct 'effective' global counter-insurgency. As has been shown in the previous chapters, the reformulation of understandings of sovereignty, territory, and territoriality in the political and legal realms, has raised questions about how 'contingent sovereignty' – particularly with regard to security concerns – is affecting both 'putative deterritorialisation' and '… the concomitant reterritorialisation of state space relations in the present era' (Elden 2006: 21–22). For example, the use of RPA strikes can encompass processes of deterritorialisation as a claim to a particular – often legally sovereign – space is usurped for the purposes of reterritorialising an area in a fashion more amenable to counter-insurgency concerns. Since, as Elden (2010: 809) notes, 'territory is dependent on a number of techniques and the law' – or practices of territoriality – the RPA is playing an important role both practically and conceptually in contemporary geopolitical understandings. This has led to three inter-related discussions that are influenced by the consequences of the RPA on the ordering of space. The first has occurred in the legal realm and raises questions noted in chapter 2, about sovereignty, preemption, and distinctions – if any – between battle-spaces and spaces of policing in the everywhere war. The second is in the contending cartographic practices of targeted killing proponents and opponents in the popular mapping of drone strikes as well as their corresponding infographics.[6] Thus, as with other cartographic practices, one can see the mutual constitution of how strikes are being recorded, ordered, and performatively positioned within narratives used to provide meaning to the everywhere war. The third has been the ways in which RPA warfare has been able to connect practical and popular geographical imaginations to the micro-level representations provided through RPA technology with an unprecedented velocity. Thus, as suggested by Paul Virilio (2002), while RPAs can collapse distances between sites of control and controllers, their seductive power also lies in the claimed ability to collapse distance in (near) real time (see also Bartram 2004). Thus, practices of control can be observed, analysed, and adjusted with a speed not previously possible at a distance from the battle-space.

The implications of drone warfare for broader geopolitical understandings of space both inside and outside of contemporary counter-insurgency have given rise to a considered, important, and growing volume of research. Yet, there has been less emphasis on exploring how RPA strikes are productive of the quotidian scales and understandings of place, that is those places that are '...invested with understandings of behavioural appropriateness, cultural expectations, and so forth [in which] we act' and which hold value for us (Harrison and Dourish 1996: 3). In order to begin to think about place and how the everywhere war is also an everyday war, I suggest that two disparate strands of research are instructive: the first is literature that explores the governing of sacred places through modern practices of war-fighting and the second is recent literature that has explored the (geo)politics of the home. My argument is that by turning to these literatures and then applying their insights to the practices of targeted killing reveals the ways in which homes are productive of this strand of counter-insurgency and the significant geopolitical implications that arise when the home becomes a prominent target for 'control without occupation' via the RPA (Weizman 2007: 239). This will provide additional insight into the territorialising processes of the targeted killing assemblage.

Sacred places and counter-insurgency

While the position of sacred places in the battle-space has a long provenance in security practices for states such as India and Israel, their significance has only recently become a prominent concern within American counter-insurgency doctrine. According to Ron E. Hassner (2006: 150) sacred places are '...sites at which the heavenly and the earthly meet, providing meaning to the faithful by metaphorically reflecting the underlying order of the world'. Sacred places have generally been understood as formal sites of worship, such as mosques or temples, and/or focal points for religious pilgrimage. Shampa Mazumdar and Sanjoy Mazumdar (2004: 385) note that conceptions of the sacredness of a place and the practices through which this status is reproduced are '…learned through processes of socialisation…' which are active and experiential. Thus, the transcendental and holy qualities of formal sacred places may also be reliant on a series of practices, rituals, configurations, and understandings that are undertaken in more mundane, though no less sacred, environments. As Lily Kong (2001: 218) argues, sacred places are ordinary places made extra-ordinary through ritual, revelation, and states of consciousness. Moreover, 'sacred place is both local and universal and therefore exerts centripetal and centrifugal forces simultaneously', by drawing in and repelling its constitutive elements (Kong 2001: 218).

My argument is that by examining the sites where drone strikes are occurring, one can see that notions of sacred place are important to understanding the cultural politics of targeted killing. My contention is that the current discussion of sacred places within counter-insurgency *writ large* is hampered by two fundamental weaknesses. The first is the structure through which sacred places are being defined. By creating a binary rather than relational distinction between sacred and profane space, a significant calamitous effect of targeted killing is overlooked. The second is that the definitional weakness occludes the ways in which the desecration of other sacred places – particularly the home – undermines the everyday spatial orders that are productive of the rhythms – and potential securities – of everyday life.

The growing profile of sacred places in the counter-insurgency wars that have taken place in Afghanistan, Iraq, Libya, Pakistan, Palestine, Somalia, Syria, and Yemen primarily reflects two concerns. The first has been the way in which insurgents have used sacred spaces – particularly mosques in Iraq – as a means of reducing disparities in strength and the reluctance to overtly target these sites for military actions by counter-insurgency forces. The second has been a strategic desire not to further inflame perceptions that counter-insurgency is a war on Islam. Thus, it has been claimed that show-ing respect for sacred sites and those who believe in them has been an objective of these interventions and occupations, with uses of force reserved for only the most special circumstances (Nagl et al. 2008). In documents made public by Wikileaks, the American rules of engagement in Iraq stated that:

> civilian structures, especially cultural and historic buildings, nonmilitary structures, civilian population centres, mosques and other religious places, hospitals and facilities displaying the red crescent or red cross, are protected structures and will not be attacked except when they are being used for military purposes.

> (US Armed Forces 2005b)

Others claim that even their use for military purposes was not necessarily enough for these places to be targeted. In his memoir of his time as a combat active drone operator, Martin alleges that various insurgent groups in Fallujah and Najaf utilised mosques as storage depots and firebases, knowing that 'nothing could be shot into the zone, even in self-defence' (Martin and Sasser 2010: 68). Describing the Iman Ali Mosque in An Najaf – at the time the stronghold of Muqtada al-Sadr and the Mahdi Army, he continues:

> I was just as frustrated as everyone else, even though I understood the Muslim veneration for the ancient shrine. It was an incredible site to behold, even from ten thousand feet above it. The twin spires and gilded dome of the magnificent edifice glistened in the sun like burnished brass.
>
> (Martin and Sasser 2010: 68–69)

These conflicting sentiments are argued to be the crux of the problem constituted by sacred places for counter-insurgency within dominant discourses. Hassner (2006: 150) has stated that contemporary irregular warfare, particularly forms that were waged in Iraq, puts '...US forces in a precarious position: choose between desecrating a sacred space or restrict their fighting to respect the opponent's religious sensibilities'. Desecration is argued to be a significant problem because:

> the transgression of the boundary between the sacred and the profane, is more than just an offence to the sensibilities of those who revere a sacred site. Believers view such an action as a tangible assault on the status of the site, that if successful, can strip it of its sanctity.
>
> (Hassner 2006: 151)

The evidence would seem to suggest that for the United States the political consequences of desecration in counter-insurgency are considered undesirable with respect to those places formally coded as sacred, such as mosques – though empirical evidence from the war in Iraq could be used to demonstrate that this was always a negotiable aversion shaped by perceptions of military necessity. Yet, what I will demonstrate in the analysis to follow is that a focus on formal sacred places occludes the ways in which homes have become a primary site for RPA strikes in the everywhere war. To support this claim, I first engage with literature that has explored the geopolitics of the home in order to situate this site conceptually within the practices of counter-insurgency. I present evidence compiled from the Bureau of Investigative Journalism database of drone strikes in Pakistan, Somalia, and Yemen to illustrate the extent to which homes are a focal point for targeted killing. Using the work of Juan E. Campo (1991), I will indicate how the familial home can be understood as a sacred place within Islamic faith traditions and the potential implications of these strikes on the local spatial order.

The geopolitics of the home

Unpacking the geopolitics of the home produced through drone warfare requires that the home itself must be conceptualised in its myriad dimensions. Brickell argues that the home should be seen as both a physical and imaginary place that 'is not separate from public, political worlds but is constituted through them: the domestic is created by the extra – domestic and vice versa' (Blunt and Dowling 2006: 27 quoted in Brickell 2012: 575). Through this relational dynamic, the forms that the home can take, their functions, and the ways in which they are understood can be diverse. Home can be a house – and this will be a focal point for the analysis that follows – a neighbourhood, a landscape, a climatic region, a socio-cultural setting, or some other config-uration that provides a sense of 'being at home'. Jacobsen (2009: 356) argues that 'being at home' is an:

> …intersubjective way of being that is marked by a sense of 'my own' or, more properly, '*our* own', an intersubjective way of being that is familiar and secure…even if this security is one of being comfortable in relation-ships and ways of behaving that are marked by great danger and instability for those involved.

Thus, the home is a place that is constitutive of the self and can potentially serve as a stable refuge from that which is seen as 'alien'; it is a site from which we engage with the world (Jacobson 2009: 357, 359, 361). It can also be a place where our rhythms and routines are established, serving as an extension of our physiological and emotional needs (Jacobson 2009: 362). As Ian Tucker (2010: 528) explains, 'homes are…spaces imbued with emotions, relations, and histories, not just spaces defined in a mundane sense according to a broad array of domestic activities'.

In all of these ways, a home works to constitute and maintain a particular spatial order that orients us, not just in time and space, but also ontologically in relation to the meanings attributed to this spatial order, locations within it, those who traverse it, the ways in which it can be traversed, and the forms of movement that are prohibited (Jacobson 2009: 369). The home is therefore a site in which broader relational forces 'from outside' enfold into the micro-spaces in which we spend so much of our time (Tucker 2010: 528). And the ways in which we understand and navigate these enclosures to feel at home are historical as well as contingent (Jacobson 2009: 372).

As was remarked in passing above, the house is one of those physical infrastructures that can contribute to a sense of home. The spatial, social, cultural, political, and affective influences of the house as home are important. As Campo (1991: 5) argues houses are more than a physical infrastructure to meet storage and shelter needs. Houses encompass ideas, images, practices, and particular institutional orders. They are constructed, not just through the harnessing of materials, but also through imaginations, discourses, and emotional attachments. As homes, houses therefore play an important role in the production of ontological security. By '...harbouring us against chaos and meaninglessness, they embody representations of relations between social and natural worlds, and they often furnish us representations for these worlds' (Campo 1991: 5). Even when the house is a space that enables heinous abuse, it still may be the case that those living under these conditions of insecurity perceive it as 'home'.

Houses, as sites of ontological security, contribute to identities – and their associated forms of security – that may be constituted, reproduced, or even contested through domestic practices and spatial relations. The home is symbolically and materially important in:

> ...shaping and reproducing the ideologies, everyday practices and material cultures of imperial power, nationalist resistance, and diasporic resettlement...[The house as] home...is intensely political, both in its internal intimacies and through its interfaces with the wider world.
>
> (Blunt 2005: 510)

Brickell (2012) identifies three ways in which the home is constitutive of geopolitics. The first is the role of the home in modern warfare, particularly through the deliberate destruction of the home, or what Porteous and Smith (2001) refer to as 'domicide' (Brickell 2012: 577). The second is the role of the home in the construction of homelands and nations. This can range from the practices of 'domopolitics' (Walters 2004), defined as 'the aspiration to run the state and homeland like a protected cocoon of community and citizenship, set against a dangerous outside of illegals, traffickers and terrorists', to functions within diasporic homes, to forms of resistance, to land grabs in the global south (Brickell 2012: 578). The third is geopolitical homes which can range from examinations of the quotidian practices of home-making that sustain armed forces and protestor encampments to those that serve as refuges from institutionalised discrimination and oppression. As Brickell (2012: 584) demonstrates through this topology, 'domestic spaces...play a critical role' in various iterations of the everywhere war as an everyday war.

One of the ways in which our houses and homes produce and perform ontological security is through the ways in which we may link these places to the transcendental. The interplay between the home and religion can be important to the ways in which people '...develop profound attachments to the places in which they reside, or in which they imagine themselves to reside' (Campo 1991: 3). In the section that follows, I examine research that has looked at the specific role of the home within societies that are predominantly Islamic in order to demonstrate the overt ways in which houses may be understood as sacred places. The point is to reveal the potential for additional injury – beyond phenomenological attachment and ontological orientation – to those who have thus far been subject to targeted killing strikes that is at odds with understandings prevalent within the US military regarding the respect shown to sacred places in the everywhere war. In the next section, I examine the conditions of possibility within Islam for understandings of the home as sacred place.

The home and Islam

In exploring the sacred relationship between Islamic faith traditions and the home, it is important to avoid Orientalist tropes that position Islam as unchanged for millennia, '...emphasize the archaic, the primitive, and the "pure...", and fail to differentiate by assuming universal beliefs and practices within and across Muslim societies (Fuchs 1998: 157). Similarly, the point is not to position Islam as abjectly different or exotic in relation to other religious or secular societies with regards to how the home is seen as both sacred and as a sanctuary from outside forces. For example, most societies (Islamic and otherwise) have clear legal rules with regards to the procedures and circumstances under which an agent of the state may enter a home uninvited. More extremely, during the troubles in Northern Ireland, doorstep killings, sectarian murders in which paramilitaries would traverse into opposition territories and ambush their targets as they came to answer the door of their houses, were seen as a particularly egregious form of political violence because these acts violated both communal and domestic spaces. The taboo against the violation of the home space was such that most paramilitaries who engaged in this form of killing preferred not to cross the threshold in pursuit of their victims (Feldman 1991: 71).

In relation to Islam, Campo (1991: 3) argues that '...ethnographic research suggests that Muslims widely attribute religious significance to their houses today, even when their dwellings happen to be apartments designed according to European and American prototypes'. This religious significance, despite Western perceptions of strict Islamic regulations that govern all aspects of behaviour, has been established without the presence of formal codes regarding domestic life arising from the Quran or key Hadiths. Thus, research has explored '...how Muslims appropriate Islamic practices and symbols in the improvisation of a domestic order that has religious meaning' (Campo 1991: 5). Campo's grand argument is that the house as a particular kind of place, links the spiritual (the House of God), the politico-economic (the House of Islam), and the everyday (the Muslim home). These interconnections are both long-standing and vital as 'the establishment of Islam is intimately connected with the creation, appropriation, and expansion of Muslim domestic space' (Campo 1991: 65). Sacredness has thus provided a tacit linkage between the human household, God's house, and the Prophet's house-mosque. And underpinning these links can be a more general politico-religious project that seeks to make the domestic order and the social order congruent.

In Islamic traditions, the Quran links domestic life, including particular practices such as hospitality, etiquette, and the spatial configuration of the home (often understood as a mixture of making visible and invisible through the organisation of space), to notions of purity required to demonstrate sub-mission to God and to be worthy of the realms of the blessed in the hereafter. The Hadiths provide a broad interpretation of what constitutes a mosque with Muslim houses being granted a special status. In turn, Campo (1991) notes that the terms *al-dar* and *bayt* – which are vernacular terms for house – are used within the Quran and the Hadiths to convey the domestic char-acteristics of key cosmological locations like heaven. It is through prayer and religious adherence that the good life may come to a home. Particular prac-tices and norms such as visitation rules also serve a higher purpose. Campo (1991) argues that these are about balancing the imperatives of being a good host with maintaining the sacredness of the home. Preserving a house as a sacred area (*haram*) also requires the respect of others and thus there are axioms that should guide the behaviour of guests.

What becomes clear from Campo's theological and ethnographic analyses is that sacredness is not innate; it becomes manifest through domestic practices which are then granted transcendent value (Campo 1991: 166). Therefore, '... features of domestic significance come more from the appropriation of key Islamic symbols by the people than from an instituted canon of formal religious requirements' (Campo 1991: 137). Moreover, religious meaning is not deter-mined by the form of the dwelling. Even the most modest house can be organised such that it becomes perceived as a sacred place by its inhabitants. Religious meaning therefore should be seen as a product of social action, speech, and interior display (Campo 1991: 8–27). In addition, while homes may have religious significance for their dwellers in predominantly Islamic societies, what makes them sacred need not be limited to a religious register. In this way, just as with the argument above regarding sacred places more generally, boundaries are important, not because they represent a fixed line of demarcation, but because of their relationality and how residents imagine that relationality. The relationality of the domestic/sacred and profane draws our attention to the politics that becomes possible when boundaries are estab-lished, maintained, and circumvented (Campo 1991: 95). In the following section, I explore the geopolitics that becomes possible when RPA strikes target houses in predominantly Islamic societies. Using the Bureau of Investigative Journalism's database on drone strikes for Pakistan, Somalia, and Yemen, I first establish the extent to which houses have featured in the commissioning of this form of political violence. I then turn to first person accounts of those who have lost their homes in order to show their own understandings of the consequences of this geopolitics.

Drone strikes and houses

In presenting an argument about the extent to which houses have featured in RPA strikes, I am not seeking to downplay the horrific and tragic human costs of deaths and injuries suffered by civilians. My focus on the targeting and destruction of houses is to demonstrate that the affective and political costs extend beyond the most obvious forms of loss and injury. Desecrating homes not only contributes to lasting socio-psychological harm but has also been suggested as a catalyst for the creation of more constricted – if not fundamentalist – everyday spatial orders that shape (geo)politics.

Data for this analysis was compiled in July 2013 from qualitative data sets on drone strikes collected by the Bureau of Investigative Journalism (BIJ) on Pakistan, Somalia, and Yemen which date back to 2002.[7] This dataset was used because it drew upon several different sources of material including official US government statements, information provided by local authorities, and on the ground investigation. Moreover, many individual cases included links to media reports, both local and international, all of which helped in interpreting the data. I began by establishing codes for potential sites that have been attacked. Eight coding categories were created. These were: 'formal religious', 'houses', 'transportation', 'commercial', 'education', 'military', 'other', and 'unknown'. Formal religious was defined as those sites of formal religious significance such as mosques. Houses were defined as primarily domestic spaces where people live. Transportation was defined both as vehicles and spaces such as bus depots. Commercial sites were understood as areas of business activity including shops and marketplaces. Education was understood to include places such as schools and vocational training centres. Military sites were defined as training camps, bases, hideouts, or other forms of infrastructure that served a direct military purpose for insurgents. Other referred to any site that did not fit into the above categories. Unknown was used to code instances where it was unclear from reports what kind of site had been subject to a targeted killing operation.

The coding did not attempt to capture intent with regards to what was destroyed. In other words, the process did not seek to discover insight into primary targets and collateral damage. Neither did it adopt dominant counter-insurgency logics that understand any site at which a militant – or suspected militant – was present as a military target. Coding also did not seek to quantify the specific number of site types destroyed in a given strike; coding for individual cases was thus a measure of presence rather than frequency. These determinations – even if desirable – were not possible from the data that are available. Rather, the aim was to account for the kinds of spaces and places that become theatres of counter-insurgency through the practice of targeted killing strikes by RPA. For example, on 10 January 2013, the small village of Hesso Khel (near Mir Ali in North Waziristan) experienced its second drone attack in three days which destroyed a house and a moving motorcycle. This strike, in which three to six people were reportedly killed and several others injured, would have been recorded as encompassing both domestic and transportation categories.

Coding was undertaken by three coders in 2014. There was an initial pilot involving one coder in July 2013. After the initial results of this pilot study looked promising, two additional coders were trained and another pilot of a small sample of cases was undertaken to ensure coding reliability. 492 individual cases were coded and each case was coded independently by each coder.[8]

ReCal (Freelon 2013) was used to calculate inter-coder reliability using percentage agreements (pairwise and overall), Fleiss' kappa, Cohen's kappa (pairwise), and Krippendorf's alpha. Kappa values were calculated to determine the level of agreement beyond that which could be obtained by chance. The alpha value was calculated to allow for potential comparison to future analyses of datasets with different coders, values, metrics, and sample sizes. All measures for the coding across the 492 cases by the three coders met the reliability thresholds established in the literature: p>0.9 overall and pairwise agreement; k>0.8 for kappa values; and α>0.8 alpha values (Krippendorff 2004: 429; Viera and Garrett 2005: 362).[9] See Table 6.1 for a listing.

Despite the overall level of coder reliability as indicated by percentage agreements, kappa values, and the alpha value, there were still some issues with conducting the coding. First, there was a lack of information for some incidents – particularly in the cases of Yemen and Somalia – about what had been hit, where a strike had taken place, and if drones – as opposed to cruise missiles, regular aircraft, or attack helicopters were involved. In these instances, caution was exercised and events were only recorded if they plausibly looked to be a drone strike from the evidence available in the archive. Second, official statements and eye witness accounts of what had been hit in a given drone strike were often at odds with one another. In these instances, coding was undertaken on the basis of the weight of the evidence, paying particular attention to local media and eye witness reports collected by the BIJ over statements from the American, Pakistani, or Yemeni governments. Third, it was important to ensure that spatial descriptors common in US counter-insurgency policy did not over-determine coding. In particular, the term 'compound' was often deployed as a means of presenting a particular space as a legitimate

Table 6.1 Measures of coding reliability

Measure	Value
Average pairwise % agreement	0.9133
Pairwise agreement coders 1 & 3	0.9207
Pairwise agreement coders 1 & 2	0.9024
Pairwise agreement coders 2 & 3	0.9167
Fleiss' kappa	0.8891
FK observed agreement	0.9133
FK expected agreement	0.2180
Average pairwise Cohen's kappa	0.8891
Pairwise CK coders 1 & 3	0.8992
Pairwise CK coders 1 & 2	0.8748
Pairwise CK colders 2 & 3	0.8933
Krippendorff's alpha	0.8892

military target because of its association within the American vernacular with dangerous spaces (e.g., armed cults like the Branch Davidians). Housing compounds made up of several buildings to accommodate civilians are not an unusual housing configuration, particularly in Pakistan. Thus care was taken to ensure that domestic housing areas were not conflated with insurgent bases to the extent possible given the quality of the available data.[10]

While the number of case incidents is relatively high, it is important to note that the conclusions that can be drawn from them are still limited. First, the analysis ends in July 2013, reflecting both the time this work was undertaken (summer 2014) and the need to draw upon a select number of cases for pilot/ training purposes. Second, the database itself only provides comprehensive longitudinal data for targeted killing operations that have taken place in Pakistan, Somalia, and Yemen.[11] For example, it does not include over 1,200 drone strikes across Afghanistan, Iraq, and Libya undertaken by the regular branches of the armed forces that have been recorded by US CENTCOM (Woods and Ross 2012). Third, the incidents contained in the database are those that have been alleged to have been undertaken by the Central Intelligence Agency as part of 'personality strike' operations that are approved by the office of the president of the United States. At the time of writing, while there are some publicly available databases of drone strikes conducted by the American and British armed forces in Afghanistan and Iraq, these are partial.[12] Given that it is conceivable that the military selects different sites to be targeted and follows different targeting procedures, it would be overstating the case to claim that the findings below fully capture the distribution of spaces targeted by drone strikes in the everywhere war. However, they are at least indicative of the prominence of the destruction of homes within targeted killing operations conducted via drones.

Findings

The number of probable RPA strikes examined was 492. The coverage spanned from 2002 to 2013 with the earliest incident taking place in Marib Province, Yemen on 3 November 2002 and the last on 3 July 2013 in North Waziristan, Pakistan. Out of the total number of probable strikes, 387 of were recorded in Pakistan, 96 in Yemen, and 9 in Somalia. Out of the total of 492 cases, there was unanimous coder agreement across all coding decisions in 432 cases and at least partial coder agreement (i.e., agreement between at least two out of three coders on a coding decision as noted above) in 58 cases. Thus, a total of 490 cases contained at least partial agreement. There were two cases where the coders came to no agreement.

Descriptive statistics reported below are based on at least partial agreement for coding, defined as two out of three coders agreeing that a particular space had been hit in a strike.[13] In order to determine if this threshold might skew results, partial agreement descriptive statistics were compared with cases of unanimous agreement overall and within individual theatres (i.e., Pakistan,

Somalia, and Yemen). Two tailed Z-tests at p>0.9 showed no statistically significant differences in results obtained through partial agreement and unanimous agreement overall and in individual theatres.

According to coding of strike incidents contained within the databases, targeted killing strikes via RPA in which at least one house was destroyed or damaged, represented 54.5 per cent of the total number of strikes undertaken during the time period examined in these three counter-insurgency theatres (see Table 6.2). There was, however, significant variation amongst the three theatres. In Somalia – with a low number of cases – there have been no reported incidents of houses being targeted (see Table 6.3). In Yemen, approximately

Table 6.2 Strike Types in Pakistan, Yemen, and Somalia as Percentage of Total Strike Events (01/01/2002 – 30/07/2013)

	At least partial agreement *(n=490)*
Type of space hit	Percentage
Domestic	53.5
Transportation	46.5
Unknown	9.1
Military	6.5
Education	4.1
Commercial	2.4
Other	1.8
Religious	1.0

Total cases coded n=492.

Table 6.3 Strike Types as a Percentage of Total Strike Events in Somali (01/01/2011 – 31/07/2013)

	At least partial agreement cases *(n=9)*
Type of space hit	
Unknown	44.4
Military	33.3
Transportation	22.2
Commercial	0
Domestic	0
Education	0
Other	0
Religious	0

Total cases coded n=9.

12.5 per cent of strikes involved hitting a house or houses (see Table 6.4). Yet, in Pakistan, the proportion of strikes that destroyed or damaged homes during the time period covered was 64.9 per cent (see Table 6.5). This is a similar result to a study by Ross and Searle (2014) using the same dataset (but a different timespan). They found that 61 per cent of strikes in Pakistan had hit a home.

To put the extent to which domestic spaces became sites of drone strikes into perspective, the analysis of the data indicated that only 6.5 per cent of attacks across all three theatres were directed at military targets such as training compounds, weapons depots, bases, or other spaces identified as sites of overt insurgent activity (see Table 6.2). This compared to approximately

Table 6.4 Strike Types as a Percentage of Total Strike Events in Yemen (01/01/ 2002 – 31/07/2013)

	At least partial agreement (n=96)
Type of space hit	Percentage
Transportation	71.9
Unknown	14.6
Domestic	12.5
Military	8.3
Commercial	2.1
Education	2.1
Other	2.1
Religious	1.0

Total cases coded n=96.

Table 6.5 Strike Types as a % of Total Strike Events in Pakistan (01/01/2004 – 31/07/ 2013)

	At least partial agreement (n=385)
Type of space hit	Percentage
Domestic	64.9
Transportation	40.6
Unknown	7.2
Military	5.4
Education	4.7
Commercial	2.6
Other	1.8
Religious	1.0

Total cases coded n=387.

46.5 per cent of strikes that involved a vehicle – a vehicle was hit in approximately 71.9 per cent of the RPA strikes in Yemen (see Table 6.4) – and roughly 2.4 per cent of RPA attacks during the time period across all three states involved a commercial property or space (see Table 6.2). Similarly, formal religious sites featured in a very small percentage of the total strikes, with these places found in approximately 1.0 per cent of the drone strikes recorded by the BIJ for which there was at least partial agreement (see Table 6.2).

The data indicated that there was no trend over time in terms of the types of sites hit by drone strikes, particularly across the three theatres involved. The predominance of any given site in any given year has experienced ebbs and flows that do not reveal specific patterns, apart from the observation noted above that houses are hit much more frequently in Pakistan than Yemen or Somalia.

Estimations of civilian casualties were not undertaken as several previous studies have attempted to quantify the immediate human toll of the drone war (e.g., Bureau of Investigative Journalism 2015; New America Foundation 2015a; 2015b). What did become clear through going through the database was that geographic locales targeted were more populated than one might otherwise think from US government and media representations of them. For example, according to census figures collected by the Pakistani government, in the Federally Administered Tribal areas (FATA) which includes North and South Waziristan, the most frequent sites of drones strikes in the area, the total population was 4,301,732 in 2012, with an average population density of 158 people per km^2 (Directorate of Health 2014: 8). By way of comparison to UK regions, FATA is more densely populated than Northumberland, Shropshire, Carlisle, Harrogate, and Cornwall. Unfortunately, more fine grained population data for towns like Datta Khel, Mir Ali, or Miran Shah that have experienced multiple drone strikes were not available.

The geopolitical consequences of the destruction of homes

From the publicly available data that has been collected, including first-hand accounts with drone strike survivors, it is clear that houses are being destroyed – and in specific cases, intentionally targeted. Some residents believe that this is part of an orchestrated campaign to displace the civilian population. For example, after a drone strike in Ja'ar Yemen in the spring of 2012 – an area that had been subject to intense fighting between government and anti-government forces – one resident disputed the claim that the house targeted was occupied by a militant commander. Rather, he insisted that the attack was part of a larger project to 'force the inhabitants to flee Ja'ar' (Alkarama Foundation 2013: 14). However, there is currently no publicly available evidence of the intent to systematically destroy homes through drone strikes, thus falling short of the definition of domicide proposed by Porteous and Smith (2001); however, the absence of evidence of domicide does not mean that consequences are not detrimental, for the targeted populations as well as in terms of the goals of counter-insurgency and transforming dangerous spaces through associated modes of governance. In particular, the destruction of homes through targeted killing operations undertaken by drones demonstrates a complete insensitivity to the economic, social, cultural, and political consequences of this form of counter-insurgency. These can be found at the individual and community levels that feed into broader geopolitical impacts.

As mentioned in the previous chapter on the aesthetic subjects of RPA warfare, those who experience strikes – either directly or indirectly – or who live where strikes are probable, suffer. People interviewed for the Stanford and NYU report 'Living Under Drones' described how 'the presence of drones and capacity of the US to strike anywhere at any time led to constant and severe fear, anxiety, and stress, especially when taken together with the inability of those on the ground to ensure their own safety' (Stanford and NYU 2012: 55). A more recent Alkarama investigation of populations experiencing drone strikes in Yemen found that 72 out of 100 interview subjects had PTSD with 27 deemed likely to have PTSD (Alkarama Foundation 2015: 12). This is not surprising when one looks at individual strike events. Eye witness accounts of a double tap strike that took place 15 May 2012 on a suspected militant's home in Ja'ar – in which 14 people died – conveyed feelings of shock, horror, and abandonment. One eye witness who suffered a stroke during the attack stated:

> the plane bombed close to my home. I heard the explosion, our house was shaken, and there were dead in the streets. I was experiencing high anxiety and had the attack. The state did not help me even though I am a single woman.
>
> (Alkarama Foundation 2013: 14)

A neighbour of the individual targeted by this drone strike described the event thus:

> I heard the detonation of the bomb and saw smoke. I rushed there in my car. Bystanders told me the house of Al-Arshani, close to mine, had been targeted. Once I arrived, I found my home in ruins. Three members of my family had been inside and one of them was injured while the other two remained unharmed. I took them to the home of a relative and returned to the scene. It was while I was arriving that a plane flew over a second time and bombed people who had been assisting the wounded from the first attack...I saw seven or eight people at least die at that moment.
>
> (Alkarama Foundation 2013: 14)

Micro-level geopolitics also are significant in the enabling of drone strikes against the home. One man from North Wazirstan recounted the following story about the day prior to his home being destroyed in a strike:

some Taliban had come to the house and asked for lunch. I feared them and was unable to stop them because all local people must offer them food. They stayed for about one hour and then left. The very next day, our house was hit...My only son Khaliq was killed.

(CIVIC 2010: 61)

Melmastia or hospitality is a significant cultural norm for Pashtuns as is the principle of *namwatey/nanawati*, which has been translated as 'to enter into the security of a house' or provide asylum. Both are argued to impose a high burden on Pashtuns to protect guests, including foreign fighters whose causes they may not support (Stanford and NYU 2012: 22–3). Thus, cultural obligations for the host that arise along the boundary of the domestic space of the home can lead to harm and injury.

Unsurprisingly, those injured in drone strikes or who had friends and/or family members die from drone strikes report symptoms of post-traumatic stress disorder and depression. A local religious scholar in Mir Ali, Pakistan complained that:

these planes had deprived the innocent tribesmen of their mental peace and badly affected their routine life. [People were] disappointed with their own government, as despite their repeated appeals it had failed to stop the attacks on the civilian areas.[14]

Similarly, in an interview with the Alkarama Foundation, an official from the Ministry of Education of Abyan Yemen lamented that houses were targeted in drone strikes. He stated:

I do not understand why they would be targeted. The consequences for the residents' peace of mind, especially the children, have been devastating for those who have experienced the trauma. The victims still have not been compensated...

(Alkarama Foundation 2013: 15)

According to research undertaken in Pakistan and Yemen, the lack of compensation for victims has amplified feelings of betrayal and helplessness that result from physical and material loss (CIVIC 2012; Amnesty International 2013). After a drone strike in Azzan, Yemen on 30 March 2012 killed three men and injured five children, as well as destroying and damaging several houses, one resident complained that the Yemeni government had failed to provide any assistance while another talked of seeking refuge with his family in a different part of the country (Alkarama Foundation 2013: 12). In Pakistan, CIVIC (2010) noted that the inability to rebuild homes destroyed in drone attacks has become a major source of grievance for those who have been affected by targeted killing operations. In some cases, it was reported that this had led to families migrating to other parts of the country. One man whose familial home was destroyed in an attack in Pakistan stated:

> I feel helpless and alone in the world. In total, 18 lakhs (approximately 21,000 USD) were spent on the construction of the house...we want the Government of Pakistan to provide us with assistance in cash to reconstruct the house. We would also accept help from the US or UN...but no one has accepted responsibility.
>
> (CIVIC 2010: 63)

Drone strikes for targeted killing not only destroy houses where high value targets may be present but can also cause damage to surrounding houses. As a result of a drone attack that took place in Kashamir on 29 August 2012, residents told human rights reporters how the strikes caused windows to break, walls to crack, and the collapse of some rooms in buildings within the immediate vicinity. Moreover, they were able to show the investigators evidence of the explosions and missile fragments that had become embedded in the walls of houses close to the site (Alkarama Foundation 2013: 16).

The economic impact of this housing damage is significant, particularly given that in the regions of Pakistan and Yemen where drone strikes take place and levels of poverty are high, home insurance and financial savings are unusual (Stanford and NYU 2012: 77). Similarly, the lack of affordable health care has allegedly led to innocent victims of strikes having to sell their houses to meet medical expenses in Yemen (Alkarama Foundation 2013: 20). Even in the absence of bodily injuries, the inability to rebuild is devastating. A 45-year-old rural farmer whose house was destroyed in a drone strike remarked:

> A drone struck my home...I [was at] work at that time, so there was nobody in my home and nobody killed...Nothing else was destroyed other than my house. I went back to see the home, but there was nothing to do – I just saw my home wrecked...I was extremely sad, because normally a house costs around 10 lakh, or 1,000,000 rupees [US $10,593], and I don't even have 5,000 rupees now [US $53]. I spent my whole life in that house...my father had lived there as well. There is a big difference between having your home and living on rent or mortgage...[I] belong to a poor family and my home has been destroyed...[and] I'm just hoping that I somehow recover financially.
>
> (Stanford and NYU 2012: 77)

Despite the wider damage to domestic infrastructure that results from drone strikes, there had been a general perception – at least within parts of the Federally Administered Tribal Areas of Pakistan – that attacks were relatively precise. For example, the International Crisis Group (2013: 25) reported that a 2009 survey of people in North and South Waziristan by the Aryana Institute for Regional Research and Advocacy revealed that 52 per cent of the respondents believed that drone strikes were accurate. By 2011, previously held beliefs about the accuracy of strikes were not translating into approval. A survey conducted by the Community Appraisal and Motivation Programme showed just 4.3 per cent support for the statement that drone strikes are 'sometimes justified, if properly targeted and excessive civilian casualties are avoided' (quoted in International Crisis Group 2013: 25). However, as outlined in the previous chapter, even if strikes are not seen as being justified, perceptions that strikes are generally accurate can make it difficult for people who have been targeted to clear their names or escape suspicion that they must have been involved in something untoward. Moreover, for those believed to have brought the attention of drones into a community, there can be resentment and a lack of sympathy for any losses incurred.

The archive collected thus far by humanitarian organisations reveals that the individual health consequences and financial implications from the loss of homes in drone strikes are being voiced and that these are largely articulated through common terms of discourse within the fields of medicine and economics. Similarly, the costs of strikes are also expressed in terms that demonstrate the way in which strikes disrupt the everyday rhythms of the cultural habitus. For example, interviewees in the 'Living Under Drones' report testified that:

> the fear of strikes undermines people's sense of safety to such an extent that it has at times affected their willingness to engage in a wide variety of activities, including social gatherings, educational and economic opportunities, funerals, and the fear has also undermined community trust.
>
> (Stanford and NYU 2012: 55)

Similarly, Alkarama (2015: 23) has reported that in Yemen, 'rumours that drones are able to see inside the houses and watch women have also spread, leading some women to live under the constant fear that even inside their homes, they are watched by American soldiers'.

Yet, the publicly released interviews that humanitarian organisations have conducted with drone strike survivors have not directly explored the potential connections between the domestic space and religious piety or how these connections may shape perceptions of dessert in times of misfortune. Whether this has been an oversight in the kinds of questions that have been asked or a conscious strategy to distance forms of suffering being expressed within a religious idiom that is often perceived suspiciously in the West – in contra-distinction to more universally recognised forms of suffering – requires further investigation. Although expressions of loss may not be overtly articulated through religious discourses in response to questions asked by interlocutors, expressions of loss have encompassed notions of spiritual disturbance that go beyond physiological diagnoses or the ledger sheet. In this regard, film-maker Madiha Tahir has remarked:

> When drone attacks destroy homes – as they often do – they erase entire family histories. Homes in this area are built over time as families grow. There may be as many as 50 members of a family living in one house. When you destroy structures like that, you not only destroy people, you also destroy their history. The rubble that's left in the wake of an attack is a living memory of what happened there. It embodies loss. The people in Waziristan have to live around this loss, near it, in it. They have to live among ghosts.
>
> (Bhasin 2013)[15]

Broader geopolitical impacts

While the previous section explored the micro-level effects of the destruction of homes in drone strikes on individuals, families, and communities in Pakistan and Yemen, these are not tightly contained in distinct spaces. Drone strikes and the destruction of homes resonate more broadly into the geopolitical dynamics underpinning the everywhere war and global counter-insurgency.

It is claimed that the destruction of homes and the lack of compensation offered by local and foreign governments have provided 'hearts and minds' opportunities for insurgent groups. For example, it has been rumoured that local groups linked to the Taliban have engaged in sporadic reconstruction efforts, particularly in instances where homes providing refuge for foreign fighters have been destroyed. Although there is no corroborating evidence to suggest that such largess is a standard operating procedure for these groups, and there is much evidence of the ways in which insurgent groups have undermined community infrastructure like schools, homes, and medical clinics, one wonders if the circulation of these anecdotes could improve the predominantly negative views held of militants by lower income groups in Pakistan (Blair et al. 2013).

The second impact has been large-scale internal displacement in Pakistan and Yemen as people – often those whose immediate places of residence have experienced drone activity – flee to areas perceived to be relatively safer from strikes. As of August 2015, it has been conservatively estimated that conflict has displaced at least 1.8 million people in Pakistan (with 1.5 million of the displaced originating from FATA) and 1.4 million in Yemen (IDMC 2015a; 2015b). Whereas data is less conclusive for Yemen, it is reported that much of the internal migration in Pakistan has been to urban centres rather than camps (IDMC 2015a). Often, this can result in households being detached from traditional support networks, leaving them vulnerable to exploitation as well as forms of communal violence – including militant violence directed at civilians which is present in urban areas in Pakistan. Internal migration due to drone strikes also places additional pressure on already over-taxed critical infrastructures while contributing to shortages of adequate housing. The Egyptian experience in the 1960s may provide some insight into the longer-term implications of these pressures.

Campo (1991: 133) argues that when Egypt was faced with a deterioration in housing beginning in the 1960s due to a shortage of homes, one of the responses was the adoption of forms of nihilism with regards to the social, the political, and the familial. More troubling, given the current context, was that another response was the rise of violent strands of militancy. In particular, the rise of militant Islam in Egypt sought to reinvigorate the religious purity of the social order by adopting domestic idioms and wished to implement a particularly narrow conceptualisation of the home which emphasised '…internal purity and non-visibility rather than hospitality and balance between visibility and non-visibility' (Campo 1991: 136). Although the stated rationale of targeted killing has been the elimination of militancy and the disruption of insurgent groups, the findings presented here, and past precedent, would suggest that it would be unwise to ignore the quotidian geopolitics that run through the home in the everywhere war.

Conclusions

This chapter has sought to reveal how the broader geopolitics that enable drone strikes for the purposes of targeted killing reverberate into the homes of those who have been defined as populations of interest for global counter-insurgency. By demonstrating the importance of the home as a site of drone warfare as well as a place of personal significance and potential sanctuary, it was argued that the common understanding of 'sacred place' in counter-insurgency doctrine is overly restrictive. Whether this definitional narrowness has been an act of commission or omission is for the reader to decide. Evidence has been presented to suggest the importance of homes as sites targeted in the CIA drone campaign since 2002, particularly in Pakistan. Moreover based on the fieldwork undertaken by several non-governmental organisations in Pakistan and Yemen, accounts that emphasise the importance of the home in the articulation of loss and grievance for those who have experienced strikes has been presented.

The significance of these findings is that they reveal another strand of culturally infused politics produced through targeted killing assemblage: the desecration of the home as place. This is accomplished not just through the overt destruction of the arranged physical material substance that comprises the home but also by erasing the relationships, traditions, and histories that flow through them. As I have argued previously, the symbolic power of the drone strike as a form of violence comes not just from '…the ability to defile but also from the atemporal character of defiled space – as Feldman notes, "defiled space never goes away"' (Feldman 1991: 67 quoted in Grayson 2012b: 125). Moreover, in the aftermath, defiled spaces and their spectres haunt the living. In this way, drone warfare and targeted killing demonstrate

'...how counter-insurgency also "seeks to violate the spatial constructs that function as armatures of the victim's social order"' (Feldman 1991: 75 quoted in Grayson 2012b: 125). The broader geopolitical impacts of disruptions to everyday spatial, spiritual, and rhythmic orders catalysed by the destruction of homes within the everywhere war may not have fully materialised, but previous experience suggests that it will certainly do nothing to pacify violent insurgency or undermine forms of extremism that foster it.

Notes

1 For example, see Bishop and Phillips (2002); Grayson (2012a; 2012b); Gregory (2011b); Williams (2010; 2011); Shaw and Akhter (2012); and Shaw (2013).

2 As Davies (forthcoming) suggests through his close reading of Henri Lefebvre, there is nothing inherently progressive about the everyday. As a site of routine, repetition, and regularity, it can be as stifling as any other site of politics. However, Davies (forthcoming) also reminds us that 'the prevailing image of everyday life... – whether it takes a "top down" view of the exercise of power at the international level over the everyday or whether it takes a "bottom up" view of the tactical capacities of "everyday" actors – retains a notion of everyday life as a reified category, a level "under" that of the international. Henri Lefebvre's dialectical conception restores a critical edge to...everyday life by showing how everyday life generates a critique of itself, of the higher activities, and of their initial, generative separation.'.

3 For example, Adey (2010); Adey et al. (2011); Anderson (2011); Bishop (2011); Gregory (2010); and Saint-Amour (2011).

4 A classic example would be Kaplan (2009).

5 With regards to security practices broadly, see Dillon (2007); Duffield (2007); Dillon and Lobo-Guerrero (2008); Dillon and Reid (2009). With respect to drones see Grayson (2012b); Gregory (2010); and Shaw (2013).

6 See for example 'Out of Sight, Out of Mind' at http://drones.pitchinteractive.com. Accessed 15 October 2015.

7 Datasets for Pakistan, Somalia, and Yemen were found at The Bureau of Investigative Journalism's Covert Drone War site: http://www.thebureauinvestigates.com/category/projects/drones.Accessed 15 October 2015.

8 My thanks to Hector Bezares-Buenrostro and Paul McFadden who served as fellow coders.

9 It is worth noting that these are taken as guidelines only. Krippendorff for example notes that the levels of reliability necessary will depend on the conclusions to be drawn and the nature of the research (i.e., life and death medical interpretations should require a higher level of coding agreement).

10 Compounds of homes provide an arrangement whereby extended families can live together and support one another. See Stanford and NYU (2012: 25).

11 At the time of writing, a dataset for Afghanistan has been added; however, coverage only extends back to January 2015.

12 See for example the partial dataset collected by Drone Wars UK at http://dronewars.net/uk—drone—strike—list—2.Accessed 15 October 2015.

13 Overall, there were only two cases where at least partial agreement was not reached for at least one coding category for a strike event.

14 See entry B46 at Bureau of Investigative Journalism, 'The Bush Years: Pakistan Strikes 2004–2009', http://www.thebureauinvestigates.com/2011/08/10/the-bush-years-2004-2009. Accessed 15 October 2015.

15 Thanks to Derek Gregory for this reference.

Bibliography

Adey, P. (2010) *Aerial Life: Space, Mobilities, Affects*, Oxford, Wiley-Blackwell.

Adey, P., Whitehead, Mark, and Williams, Alison J. (2011) 'Air Target: Distance, Reach and the Politics of Verticality', *Theory, Culture, & Society*, 28(7–8), 173–187.

Agamben, G. (1998) *Homo Sacer: Sovereign Power and Bare Life*, Stanford, Stanford University Press.

Agnew, J. (1994) 'The Territorial Trap: The Geographical Assumptions of International Relations Theory', *Review of International Political Economy*, 1(1), 53–80.

AlkaramaFoundation (2013) *The United States' War on Yemen: Drone Attacks (Report Submitted to the Special Rapporteur on the Promotion and Protection of Human Rights and Fundamental Freedoms while Countering Terror)*, Geneva, Switzerland: Alkarama.

AlkaramaFoundation (2015) *Traumatising Skies: US Drone Operations and Post-Traumatic Stress Disorder Among Civilians in Yemen*, Geneva, Switzerland: Alkarama.

Amnesty International (2013) *Will I Be Next? US Drone Strikes In Pakistan*. London: Amnesty International Publications. Available from: https://www.amnestyusa.org/sites/default/files/asa330132013en.pdf. Accessed 15 October 2015.

Anderson, B. (2011) 'Facing the Future Enemy: US Counterinsurgency Doctrine and the Pre-insurgent', *Theory, Culture, & Society*, 28(7–8), 216–240.

Bartram, R. (2004) 'Visuality, Dromology, and Time Compression: Paul Virilio's New Occularcentrism', *Time and Society*, 13(2/3), 285–300.

Bhasin, T. (2013) 'The Drone Effect: Ghosts, Rubble and Loss in Waziristan'. *The Sunday Guardian*. 13 August. Available from: http://www.sunday-guardian.com/artbeat/the-drone-effect-ghosts-rubble-and-loss-in-waziristan. Accessed 15 October 2015.

Bishop, R. (2011) 'Project "Transparent Earth" and the Autoscopy of Aerial Targeting: The Visual Geopolitics of the Underground', *Theory, Culture, & Society*, 28(7–8), 270–286.

BishopR. and PhilipsJ. (2002) 'Sighted Weapons and Modernist Opacity: Aesthetics, Poetics, Prosthetics', *boundary 2*, 29(2), 157–179.

Blair, Christine G.Fair, C., Malhotra, N. and Shapiro, J. N. (2013) 'Poverty and Support for Militant Politics: Evidence from Pakistan', *American Journal of Political Science*, 57(1), 30–48.

Blunt, A. (2005) 'Cultural Geography: Cultural Geographies of Home', *Progress in Human Geography*, 29(4), 505–515.

Blunt, A. and Dowling, R. (2006) *Home*, New York, Routledge.

Blunt, A. and Varley, A. (2004) 'Geographies of Home', *Cultural Geographies*, 11(1), 3–6.

Bousquet, A. (2009) *The Scientific Way of Warfare: Order and Chaos on the Battlefields of Modernity*, London, Hurst and Company.

Brickell, K. (2012) 'Geopolitics of Home', *Geography Compass*, 6(10), 575–588.

Bureau of Investigative Journalism (2015) 'Get the Data: Drone Wars', *Bureau of Investigative Journalism*. Available from: https://www.thebureauinvestigates.com/category/projects/drones/drones-graphs. Accessed 15/ October 2015.

Campbell, D. (1998) *Writing Security: United States Foreign Policy and the Politics of Identity*, Minneapolis, University of Minnesota Press.

Campo, J. E. (1991) *The Other Side of Paradise: Explorations Into the Religious Meaning of Domestic Space in Islam*, Columbia, University of South Carolina Press.

Campaign for Innocent Civilians in Conflict [CIVIC] (2010) *Civilians in Armed Conflict: Civilian Harm and Conflict in Northwest Pakistan.* Available from: http://civilia nsinconflict.org/uploads/files/publications/civilian_harm_in_nw_pakistan_oct_2010. pdf. Accessed 15 October 2015.

Campaign for Innocent Civilians in Conflict [CIVIC] (2012) *The Civilian Impact of Drones: Unexamined Costs, Unanswered Questions.* Available from: http://civiliansin conflict.org/.../The_Civilian_Impact_of_Drones_w_cover.pdf. Accessed 15 October 2015.

Davies, M. (2016). 'Everyday Life as Critique: Revisiting the Everyday in IPE with Henri Lefebvre and Postcolonialism'. *International Political Sociology*, Advance Access. DOI: http://dx.doi.org/10.1093/ips/olv006.

Dillon, M. (2007) 'Governing through Contingency: The Security of Biopolitical Governance', *Political Geography*, 26(1), 41–47.

Dillon, M. and Lobo-Guerrero, L. (2008) 'Biopolitics of Security in the 21st Century', *Review of International Studies*, 34(2), 265–292.

Dillon, M. and Reid, J. (2009) *The Liberal Way of Warfare: Killing to Make Life Live*, London, Routledge.

Directorate of Health Services (2014) *Comprehensive Multi Year Plan (2014–2018).* FATA Civil Secretariat, Available from: http://goo.gl/nTgMSC. Accessed 15 October 2015.

Duffield, M. (2007) *Development, Security, and Unending War: Governing the World of Peoples*, Cambridge, Polity.

Elden, S. (2006) 'Contingent Sovereignty, Territorial Integrity, and the Sanctity of Borders', *SAIS Review*, 26(1), 11–24.

Elden, S. (2010) 'Land, Terrain, Territory', *Progress in Human Geography*, 34(6), 799–817.

Elden, S. (2013) 'Secure the Volume: Vertical Geopolitics and the Depth of Power', *Political Geography*, 34(1), 35–51.

Feldman, A. (1991) *Formations of Violence: The Narrative of the Body and Political Terror in Northern Ireland*, Chicago, University of Chicago Press.

Foucault, M. (2009). *Security, Territory, Population: Lectures at the Collège de France 1977–1978* (G. Burchell, Trans.), Basingstoke: Palgrave Macmillan.

Freelon, D. (2013) 'ReCal OIR: Ordinal, Interval, and Ratio Intercoder Reliability as a Web Service', *International Journal of Internet Science*, 8(1), 10–16.

Fuchs, R. (1998) 'The Palestinian Arab House and the Islamic "Primitive Hut"', *Muqarnas*, 15(1), 157–177.

Graham, S. (2004) 'Vertical Geopolitics: Baghdad and After', *Antipode*, 36(1), 12–23.

Graham, S. (2011) *Cities under Siege: The New Military Urbanism*, London, Verso Books.

Grayson, K. (2012a) 'The Ambivalence of Assassination: Biopolitics, Culture, and Political Violence', *Security Dialogue*, 44(1), 25–41.

Grayson, K. (2012b) 'Six Theses on Targeted Killing', *Politics*, 32(2), 120–128.

Gregory, D. (2010) 'War and Peace', *Transactions of the Institute British Geographers*, 35(2), 154–186.

Gregory, D. (2011a) 'The Everywhere War', *The Geographical Journal*, 177(3), 238–250.

Gregory, D. (2011b) 'From a View to a Kill: Drones and Late Modern War', *Theory, Culture, & Society*, 28(7–8), 188–215.

Harrison, S. and Dourish, P. (1996). 'Re-place-ing Space: The Roles of Place and Space in Collaborative Systems'. *Proceedings of the 1996 ACM Conference on Computer Supported Cooperative Work*, New York, ACM, 67–76.

Hassner, R. E. (2006) 'Fighting Insurgency on Sacred Ground', *The Washington Quarterly*, 29(2), 149–166.

[IDMC] Internal Displacement Monitoring Center (2015a) 'Pakistan IDP Figures Analysis'. *IDMC*. Available from: http://www.internal-displacement.org/south-a nd-south-east-asia/pakistan/figures-analysis. Accessed 15 October 2015.

[IDMC] Internal Displacement Monitoring Center (2015b) 'Yemen IDP Figures Analysis'. *IDMC*. Available from: http://www.internal-displacement.org/middle-ea st-and-north-africa/yemen/figures-analysis. Accessed 15 October 2015.

International Crisis Group (2013) *Drones: Myth and Reality in Pakistan*, Brussels, Belgium: International Crisis Group.

[Stanford and NYU] International Human Rights and Conflict Resolution Clinic Stanford Law School and NYU School of Law Global Justice Clinic (2012) 'Living Under Drones: Death Injury, and Trauma to Civilians from US Drone Practices in Pakistan'. Available from: http://www.livingunderdrones.org/wp-content/uploads/ 2013/10/Stanford-NYU-Living-Under-Drones.pdf. Accessed 5 February 2014.

Jacobson, K. (2009) 'A Developed Nature: A Phenomenological Account of the Experience of Home', *Continental Philosophy Review*, 42(3), 355–373.

Jones, C. A. (2015) 'Lawfare and the Juridification of Late Modern War', *Progress in Human Geography*, DOI: 0309132515572270.

Kaplan, R. D. (2009) 'The Revenge of Geography', *Foreign Policy*, 172, 96–105.

Kong, L. (2001) 'Mapping "New" Geographies of Religion: Politics and Poetics in Modernity', *Progress in Human Geography*, 25(2), 211–233.

Krippendorff, K. (2004) 'Reliability in Content Analysis', *Human Communication Research*, 30(3), 411–433.

Martin, M. and Sasser, C. (2010) *Predator: The Remote-Control Air War over Iraq and Afghanistan: A Pilot's Story*, Minneapolis, Zenith Books.

Mazumdar, S. and Mazumdar, S. (2004) 'Religion and Place Attachment: A Study of Sacred Places', *Journal of Environmental Psychology*, 24(3), 385–397.

Mitchell, K. (2010) 'Ungoverned Space: Global Security and the Geopolitics of Broken Windows', *Political Geography*, 29(5), 289–297.

Morrissey, J. (2011) 'Liberal Lawfare and Biopolitics: US Juridical Warfare in the War on Terror', *Geopolitics*, 16(2), 280–305.

Mulvany, L. (2012) 'Pentagon Challenges Chinese Monopoly on Rare Earths'. *Bloomberg*. 7 November. Available from: http://www.bloomberg.com/news/2012-11-07/penta gon-challenges-chinese-monopoly-on-rare-earths-commodities.html. Accessed 15 October 2015.

Nagl, J. A., Amos, J. F., Sewall, S. and Petraeus, D. H. (2008) *The US Army/ Marine Corps Counterinsurgency Field Manual*, Chicago, University of Chicago Press.

Neocleous, M. (2013) 'The Dream of Pacification: Accumulation, Class War, and the Hunt', *Socialist Studies/Études socialistes*, 9(2), 7–31.

New America Foundation (2015a) *Drone Wars Pakistan: Analysis*. Available from: http:// securitydata.newamerica.net/drones/pakistan-analysis.html. Accessed 15 October 2015.

New America Foundation (2015b) *Drone Wars Yemen: Analysis*. Available from: http://securitydata.newamerica.net/drones/yemen-analysis.html. Accessed 15 October 2015.

Porteous, D. and Smith, S. E. (2001) *Domicide: The Global Destruction of Home*, Montreal, McGill-Queen's Press-MQUP.

Ross, A. K. and Searle, J. (2014) 'Most US Drone Strikes in Pakistan Attack Houses'. *Bureau of Investigative Journalism*. 23 May. Available from: https://www.thebureauin vestigates.com/2014/05/23/most-us-drone-strikes-in-pakistan-attack-houses. Accessed 15 October 2015.

Saint-Amour, P. K. (2011) 'Applied Modernism: Military and Civilian Uses of the Aerial Photomosiac', *Theory, Culture, & Society*, 28(7–8), 241–269.

Shaw, I. G. R. (2013) 'Predator Empire: The Geopolitics of US Drone Warfare', *Geopolitics*, 18(3), 536–559.

Shaw, I. G. R. and Akhter, M. (2012) 'The Unbearable Humanness of Drone Warfare in FATA, Pakistan', *Antipode*, 44(4), 1490–1509.

Springer, S. (2011) 'Violence Sits in Places? Cultural Practice, Neoliberal Rationalism, and Virulent Imaginative Geographies', *Political Geography*, 30(2), 90–98.

Strawser, B. J. (ed.) (2013) *Killing by Remote Control: The Ethics of an Unmanned Military*, Oxford, Oxford University Press.

Tucker, I. (2010) 'Everyday Spaces of Mental Distress: The Spatial Habituation of Home', *Environment and Planning D: Society and Space*, 28(3), 526–538.

United States Armed Forces (2005a) 'Kill Box: Multi-Service Tactics, Techniques, and Procedures for Kill Box Deployment' (FM 3–09.34 MCRP 3–25H NTTP 3–09.2.1 AFTTP(I) 3–2.59). Available from: https://publicintelligence.net/fm-3-09-34-kill-box-tactics-and-multiservice-procedures. Accessed 15 October 2015.

United States Armed Forces (2005b) 'Rules of Engagement for Iraq' (Annex E (Con-solidated ROE) to 3–187 FRAGO 02, OPORD 02–005), Available from: https://wikileaks.org/wiki/US_Rules_of_Engagement_for_Iraq. Accessed 5/ February 2015.

Viera, A. J. and Garrett, J. M. (2005) 'Understanding Interobserver Agreement: The Kappa Statistic', *Family Medicine*, 37(5), 360–363.

Virilio, P. (2002) *Desert Screen: War at the Speed of Light*, London, Continuum.

Walters, W. (2004) 'Secure Borders, Safe Haven, Domopolitics', *Citizenship Studies*, 8(3), 237–260.

Weizman, E. (2007) *Hollow Land: Israel's Architecture of Occupation*, London, Verso Books.

Williams, A. J. (2010) 'A Crisis in Aerial Sovereignty: Considering the Implications of Recent Military Violations of National Airspace', *Area*, 42(1), 51–59.

Williams, A. J. (2011) 'Enabling Persistent Presence? Performing the Embodied Geopolitics of Unmanned Aerial Vehicle Assemblage', *Political Geography*, 30(7), 381–390.

Woods, C. and Ross, A. K. (2012) 'Revealed: US and Britain Launched 1,200 Drone Strikes in Recent Wars'. *Bureau for Investigative Journalism*. Available from: http://www.thebureauinvestigates.com/2012/12/04/revealed-us-and-britain-launched-1200-drone-strikes-in-recent-wars. Accessed 15 October 2015.

7 Concluding remarks on the cultural politics of targeted killing

Introduction

The overall aim of the preceding analysis has been to determine how targeted killing has become possible in contemporary counter-insurgency operations undertaken by liberal regimes. I have used a variant of the assemblage concept. It has been a conceptualisation that has emphasised relations amongst component parts, the incorporation of disparate elements including the non-human, power-relations, plasticity, and the importance of discourse. In doing so, I have shown how culture has intervened through processes of coding and recoding to produce problematisations central to the targeted killing assemblage. These cultural interventions are important because they provide additional resources – imaginative, discursive, and material – necessary for the recognition of targeted killing as a legitimate and effective security practice compatible with the legal, political, moral, and economic underpinnings of liberalism itself.

In this conclusion, I will provide a broad overview of key findings from each chapter, including core elements of the targeted killing assemblage that are culturally situated. The discussion will then move to the implications of these findings, both for understanding targeted killing's conditions of possibility, as well as processes of territorialisation and deterritorialisation within the targeted killing assemblage itself. The broader significance of these implications will be related to contemporary social theories of security and violence. Primarily, I reiterate the central contribution of the book, that is, the importance of cultural mediation to liberal forms of rule. Cultural mediation is important to the ways in which liberal forms of rule are understood, their underlying logics, their (violent) practices, and the ways they should be analysed. By recognising cultural mediation and not simply reducing it to an intervening variable, international relations, security studies, and political geography will be better able to capture how forms of violence emerge from very diverse sets of formations – such as religious narratives and scientific discourses – and become entrenched in liberal regimes. At the same time, taking culture seriously enables our understandings of violence to remain sensitive to the influence of contingencies like the predominance of particular ways of seeing that emerge and are reproduced through cultural mediums like videogames. The chapter will then conclude with a final assessment of the cultural politics of the targeted killing assemblage.

Chapter overview

The central argument of the book is that targeted killing – and the recent turn to drone warfare – has been made possible through a cultural politics constituted by common values, practices, norms, understandings, sensibilities, and modes of interpretation that can be found in contemporary liberal regimes. The analysis has identified the contingently obligatory relationships amongst cultural, technological, governmental, embodied, and geostrategic components within the targeted killing assemblage. It has also emphasised the role of key pro-blematisations to the stability of this assemblage. The central argument has been underpinned by the claim that particular security problematisations central to counter-insurgency such as 'what are we allowed to do, whom are we justified doing it to, how can control be best maintained, and under what conditions of insecurity do these actions become legitimate?' are more than the sum of biopolitical rationalities, material capabilities, and/or constellations formed by sovereignty and the law; their conditions of possibility arise given the presence of cultural factors that then shape and reshape relations within the assemblage. While the importance of culture is central to this account, it is *not* a claim for cultural determinism. Rather, the point is that if we are to understand what has made targeted killing by liberal regimes possible, it is important to look more broadly at myriad cultural practices that make it sensible, that is, those practices that increase the probability of targeted killing registering with existing ways of experiencing the world and being understood as a part of the common sense of security thinking.

To these ends, the second chapter argued against critical attempts to link targeted killing to the exception through normative suspensions or the sovereign ban. It demonstrated that casting assassination as targeted killing's '*other*' opens up space for the commissioning of the latter as a legal form of state-initiated violence. While the analysis of international humanitarian law and international human rights law revealed numerous points of in-determinacy, this flexibility is not equivalent to ignoring the law, declaring that the law is suspended, or incorporating the arbitrary suspension of rules within rules. Rather it was shown that there always exists the possibility to frame plausible, though per-haps not necessarily convincing, arguments about the legality specific targeted killings both pre and post event. Keeping a rigid delineation between assas-sination and targeted killing as well as mobilising legal expertise to harness the inherent flexibility under international humanitarian and international human rights law have been adopted by Israel, the United States, and most recently the United Kingdom, in defending their targeted killing programmes *writ large*. [1] Moreover, their actions, whether initially understood as legal violations or not, expand the frontiers of what becomes understood as legally permissible. As such, via Derrida (1989/1990), the conclusion reached was that the law is both a progenitor and off-spring of political violence in general, and targeted killing in particular. Thus, the law is a stabilising influence on the targeted killing assemblage, attempting to homogenise its legal foundations through incremental processes that expand the borders of permissibility while, at the same time, sharpening its boundaries in relation to assassination to preserve targeted killing's claim to legitimacy.

The third chapter argued that politically important questions are missed when targeted killing is understood simply as a product of biopolitical rationalisations channelled through the cynegetic power-relations of manhunting. It demonstrated that cultural narratives of assassination have positioned this act of violence within an imaginary that understands it as an event with specific moral problematics and gender relations. Primarily, assassination has been understood as treacherous and therefore gendered as feminine. While presented as assassination's 'other' in legal and political discourses, targeted killing is still unable to free itself from assassination's gravitation pull. Thus, targeted killing events continue to be imbricated in a politics of meaning that seeks to manage the 'will to revenge' with considerations of who can be eliminated in this manner. The ongoing ambivalence over what constitute legitimate events and what one becomes when engaging in assassination and targeted killing, as demonstrated by the cases of Mahmoud al-Mabhouh and Osama Bin Laden, produces spaces for contestation. Thus, the turn to targeted killing by Remotely Piloted Aircraft (RPA) and the invocation of values emerging from contemporary political economy can be seen as potential counters to the destabilising potentials of this narrative ambivalence.

In the fourth chapter, it was argued that the recent predominance of the drone in counter-insurgency and its use for targeted killing must be attributed to a complex interplay of science, cultural values constitutive of contemporary capitalist relations, and war-fighting doctrines. It is not just technological development alone that has enabled the institutionalisation of the RPA; rather it is the styles of thought and their preferred organisational structures derived from contemporary science, economic thinking, and military doctrine that have given rise to the drone. Primarily, chaoplexic thinking, network centric imaginaries, and preferences for speed, maximising information flows, flexibility, delayered organisational forms, and automation have facilitated the incorporation of the drone into counter-insurgency thinking and practice. Thus, a contingent set of understandings and beliefs from a seemingly diverse set of fields has resonated to make RPAs, and their use for targeted killing, appear as common sense. More broadly, the chapter demonstrated why it is important not to divorce war-fighting doctrines from the economic understandings in which they are embedded. To do so is to miss how logics, rationalisations, and problematisations circulate and become embedded across disparate fields of expertise, creating a powerful discursive formation that stokes the need for drones amongst a range of security actors, from militaries, to local police forces.

Chapter 5 outlined the complex aesthetic subjects produced by the targeted killing assemblage including operators, targeted populations, and RPAs themselves. It identified control as a key element of how drone strikes are problematised by the US military. Yet, contra instrumental accounts of violence that assume agents and recipients are arithmetic subjects whose behaviours can be predicted and shaped, I argued that targeted killing involves affective and embodied experiences such as anxiety, excitement, and even boredom, that cannot be fully captured or disciplined. Moreover, I rebuked claims that targeted killing is enabled by clinical detachment. Instead, it was shown that proximity and attachment are produced in operators through visualising technologies like the cameras that help to form the RPA as a 'sensor-shooter' within a global 'kill chain' (USAF 2007; US Department of Defence 2009). These technologies are in turn routed through physical senses and sensory regimes that are culturally mediated. As such, it was argued that the inevitable space that exists between what is experienced in drone warfare by operators and those subject to the drone gaze, as well as how this experience is represented, is also a key battle-ground in contemporary counter-insurgency. As such, how this space is managed by authorities and participants is a vital element within the targeted killing assemblage.

The sixth chapter argued that it is important to examine the quotidian geopolitical places that make the everywhere war into an 'everyday war' and how targeted killing contributes to these processes (Gregory 2011). It showed how drone strikes that emerge from the targeted killing assemblage contribute to processes that affect quotidian spatial relations. The analysis also revealed how the expanding battle-space mobilised through the targeted killing assemblage colonises places and seeks to disrupt their temporalities. The specific focus was on the home. It was shown how the home may be considered as a sacred place, both philosophically but also through lived experiences in 'Islamic cultural zones' (Pasha 2005: 545). Content analysis of the sites of strikes was undertaken using the Bureau of Investigative Journalism's drone strike database. This provided evidence of the predominance of the home as a site for targeted killings by the Central Intelligence Agency. Thus, in desecrating the home as sacred place, targeted killing disrupts the relationships, traditions, and histories that flow through them, producing another important strand of culturally infused politics within the assemblage. These strands will be outlined in more detail in the next section.

The cultural politics of targeted killing

The proceeding analysis has demonstrated that the turn to targeted killing and to the drone as a means of delivery cannot be reduced to technological developments, the perversion of liberal forms of rule, or rationally derived responses to the innately dangerous nature of an insurgent threat to world order. Rather, it has been demonstrated that the recourse to targeted killing by liberal regimes, the ways in which it is understood, the form that debates take, values that are expressed, the ways in which authorities attempt to control this form of political violence, the people and places that are targeted, and the predominance of the drone are produced by – and productive of – culture. With regards to targeted killing, liberal political culture emphasises and navigates the following six cultural components.

The first is narratives that position acts of violence as types of events based on both the act itself and perceptions regarding the legitimacy of its constituent elements. As the social history of representations of Judith of Bethulia presented in chapter 3 attests, if stakes are considered to be particularly high, acts comprised of elements perceived to be potentially illegitimate, such as treachery or deception, may still be viewed as honourable. However, the ways in which acts of violence are gendered and ambiguity over what it means to be one who has committed such an act create anxiety. Despite efforts to dissolve the association, the narrative norms and understandings of assassination have been grafted onto targeted killing, even if it does not necessarily involve the exact forms of treachery and perfidy said to constitute assassination events. For example, drone strikes could be seen as bombing or any other form of killing in war. Yet, they continue to be positioned in relation to assassination. This ongoing connection speaks to how individualised killing of named targets is unable to shed ways it has been previously represented within what are now liberal political cultures. As the cases of Mahmoud al-Mabhouh and Osama Bin Laden suggest, even when those killed are likely to be understood by an audience as deserving of death, how such acts are commissioned and their logistical components ultimately shape how these events are evaluated.

The second have been specific sensory regimes and the ways in which these interact with technology to produce enhanced forms of perception. Within the targeted killing assemblage, it has been 'ways of seeing' that have been most prominent (Berger 2008). This reflects a long desire in Western artistic, scientific, and cartographic traditions since the Renaissance for sight through a gaze that 'sees everything from nowhere'. Currently, new technologies like Gorgon Stare and Argus – a collection of cameras that have the capability to expand the visual field captured to 10 km by 10 km areas while still offering the possibility to localise viewing down to the movements of individual human beings – have stoked the belief in the potential to see everything as it really is from an unobstructed vantage point, without interruption. It is the desire for an unwavering gaze, that is a gaze without 'blinks', a persistent form of seeing that observes for the purposes of ordering territory. It is also a form of seeing that is asymmetrical, creating a global hierarchy whose core division is between those who watch and those who are watched. As demonstrated by accounts of drone operators, what becomes novel at this stage in time is the default medium through which the battle-space and war-fighting are interpreted and understood. While Joanna Bourke (1999) has suggested that wars of the twentieth century were mediated by soldiers through expectations derived from photography and film, it is the visual practices of video games that have become the predominant frame for drone warfare. But rather than contribute to forms of desensitisation said to arise through the 'playstation mentality', it is the juxtaposition of the tactical overview of the battle-space offered by the RPA which incorporates conscious processing and intuitive awareness with the visceral intimacy of the drone's eye view that is shaping how targeted killing is experienced by its practitioners. Pilots, despite the physical distance, are immersed into their battle-spaces, an immersion that spans from recognising local patterns of life to the visual identification of the dead and dying.

The third have been the positive values attached to information, speed, flexibility, automation, and control. Their emergence as organising principles has occurred through reinforcing dynamics forged by scientific discoveries and capitalist values instantiated into practice. Again, as hard as it might be to imagine otherwise, there is no inevitability to the predominance of big data, speed, or the centralised exercise of control in delayered organisational forms. As argued in chapters 1 and 4, while the appeal of RPAs and targeted killing are based on developments in scientific theory, their transposition into other domains – and the appropriateness and accuracy of these transpositions – has been facilitated by the current cultural predisposition towards market centred thinking and its associated problematisations that focus on managing 'chaoplexity' through information gathering, speed, and adaptability. Thus, at a time when the network has become the organising metaphor shaping perceptions of security (i.e., the desirability of adopting networked forms) and insecurity (i.e., the imperative to reveal and disrupt networks that may challenge your own), the drone is perceived as both a key node for an effective global kill chain network and a means by which to find, track, target, and eliminate essential nodes in clandestine networks whatever their geographical scale (e.g., US Department of Defence 2009). More importantly, the materialisation of these logics in the drone serves an important inter-subjective function. Possessing advanced drone technology not only affirms technological superiority but also suggests attempts to assume a subject position defined by strategic sophistication and military prowess. Therefore, drone technology is proliferating from militaries to police forces as both structures are now infused with similar mixtures of market logics, network thinking, and *raison d'etre*: the pacification of problem populations (Kienscherf 2011).

The fourth has been ways in which understandings of assassination and targeted killing, as well as their instruments and agents, are gendered. Assassination itself is in part legally defined and commonly understood as being enabled by forms of deception that are treacherous or perfidious. Assassins trick their victims, betray their confidence, or attack them outside of the normal spaces of combat in ways that are at odds with the norms of militarised masculinity. It is a practice that avoids the possibility for a situation of potential combat to be recognised by all combatants and is often thought of as encompassing weakness – sometimes understood as femininity – on the part of the perpetrator. It is a form of violence where the perpetrator seeks to mitigate any risks to the self by any means available. Thus, assassination is a form of death whose tactics betray the importance of the intended target; that is, it is a form of violence that is deliberately unbecoming to its victim. Such understandings of assassination events have been culturally (re)produced through narratives, representations of iconic assassins like Judith of Bethulia, and problematisations centred on what is morally permissible to do to others under circumstances of existential crisis that have circulated for centuries. This cultural constellation that identifies particular events as assassination events has continued to exert a strong influence over how targeted killing is understood, despite efforts to distinguish the two practices and moves towards new delivery mechanisms like drones. As evidenced by controversies over their eligibility for medals, forms of training, and promotion prospects, drone operators themselves are often vilified both inside and outside of the services as failing to embody the combat risks said to identify real warriors. Chapter 5 revealed that one of the interesting responses to this cultural politics of gender has been the gendering of the drone itself with both hyper-masculine representations of its lethality juxtaposed with feminine ideas of care and protection for soldiers on the ground.

The fifth has been the predisposition to see the law as an appropriate ground to evaluate the legitimacy of targeted killing. On the one hand, this has resulted in key states such as Israel, the United States – and more recently the United Kingdom – framing cases for targeted killing under legal precepts that are claimed to permit the act under certain conditions so long as a set of procedural formalities are undertaken (e.g., is it possible to arrest the individual being targeted or will the strike meet criteria for proportionality). On the other hand, it has also led to critiques of targeted killing that share the similar assumption that the law is a constraint rather than an enabler of violence. Thus, a common line of critique against targeted killing in general, and drone strikes in particular, has been that these are by definition illegal under specific legal provisions found in humanitarian and human rights law. Such claims involve conceptual, empirical, and ideological elements. Conceptually, it accepts liberalism's problematisation of prudence, whereby the key challenge is to apply law in order to restrain sovereign power that may unjustly impede upon negative freedoms. Empirically, it shifts ethico-political questions about what is possible to do to others and reframes them as technical questions that can be objectively answered by turning to the law as a neutral arbitrator. Ideologically, the constraining view of the law performs three important functions. First, as implied above, it privileges the law as the body best suited to determining questions of permissibility. Second, by accepting liberalism's own self-understanding of the relationship between law and sovereign power more broadly, it is tactically weak by seeking salvation in a discursive formation that is unlikely to deliver protections as expected. The problem then becomes that the standard operating procedures of the law within liberal political structures are misdiagnosed as abnormal manifestations. This misdiagnosis further legitimates a particular narrative of liberalism in which it occupies a privileged position by governing through restraint. Third, the view of law in these critiques implies a legal positivism in which the law presents itself to the world in an unmediated way. The law is thus the law, full-stop. Such a view misses that the law is a dynamic substance that is contextual in its meaning and application. Similarly, legal positivism cannot account for the fact that the law is always interpreted and thus always contestable. Moreover, in relation to targeted killing, potential violations of the law actually expand the future boundaries of the law by establishing legal precedent through a lack of prosecution. It is of little wonder then that liberal regimes have used the law to justify their targeted killing programmes. As such, the law may constrain violence in certain instances, but it also provides space for its commissioning (without liability if procedural pro forma have been followed). It is therefore imperative to examine how the law works in relation to types of violence in general and specific cases in which these types of violence have been applied.

The sixth has been understandings about what is permissible to do to those identified as 'other' as well as what one may become by commissioning such acts of violence against others. Historically, a disproportionate amount of applied innovation in forms of governmentality under liberalism has been directed at the margins within territory, or, oriented towards its periphery. Whether perceived margins and peripheries have been defined in terms of race, class, gender, geographic location, or some other marker of difference, liberalism has had a particular preoccupation with these spaces – and those who live within them – as requiring vigilant oversight. While liberalism may be the best of a bad bunch for those recognised as its legitimate subjects – or with the potential to be its subjects – from the colonial wars of the past to the colonial wars of the present, it has also shown great brutality to those it identifies as being beyond reclamation. With regards to the permissibility of targeted killing, similar to previous examinations of historical violence, its probable use increases when managing groups who are perceived to be radically different from the self. Moreover, like other forms of violence, its use in counter-insurgency reflects asymmetrical relations of power. For example, in his examination of the chemical weapons taboo, Richard Price (1997: 107) notes that the use of chemical weapons during the inter-war period by Italy in Abyssinia, the UK in Palestine (and alleged deployment in India and Afghanistan), the Spanish and French in Morocco, and Japan in China reflected the idea that particular forms of violence are '...often prohibited only among "civilised" peoples, while their use against uncivilised others is tolerated. Furthermore...such weapons are only employed when an adversary cannot reply in kind'. While this example includes regimes that at the time did not understand themselves to be liberal, it indicates that differences between liberal regimes and other types when managing their others, particularly in contexts of imperialism, pacification efforts, or colonial war, are marginal. Given current threat constructions in security discourses that focus on Islamic extremism and rely on implicit allusions to standards of civilisation discourses and Orientalist geographical imaginations, it is perhaps not surprising that targeted killing by liberal regimes is being reserved exclusively for operations in Islamic cultural zones and against those identified as Muslim.[2] As such, understandings of the Muslim other as dangerous, threatening, and ungovernable can help to justify the resort to targeted killing; however, as shown in chapter 3, public justifications continue to be offered lest audiences perceive the tactic to be so suspect that it dislodges liberal regimes from the moral high ground in global counter-insurgency.

It is important to note that while some of these six cultural factors are directly related to liberalism (e.g., economic logics, the privileging of law, and representations of the other), others do not necessarily derive directly from liberalism itself, though they have co-existed with it. Thus, the targeted killing assemblage is emblematic of a broader cultural context of the present, one whose most defining feature is its ability to divorce its myriad violences committed upon others from its central operating structures. The next section will explore the scholarly and political implications of the preceding analysis and this divorce.

Implications

My analysis in this book has implications for understanding targeted killing and drones as well as more general contributions for conceptualising contemporary security practices and political violence. The first implication is that there is need for critical work on security and political violence to step away from critiquing liberalism as it understands itself and instead critique liberalism as a violent (and often conflicting) set of practices. For defenders of the status quo, that liberalism may be less violent than other forms of political organisation (for some), should not be the evaluative standard; it is still violent. But in critiquing the violent practices – like targeted killing – undertaken under liberalism, it is important that the identification of exceptionalism as a 'cause' ceases as a line of argument. As the analysis above has shown, the concept is not particularly helpful in delineating what enables forms of political violence under liberalism. The second implication is that it is worth keeping in mind that liberalism as a form of biopolitics is not always what liberalism or biopolitics understands itself to be or what it is doing. What enables liberal biopolitics is a complex set of understandings, logics, and material capacities that extends beyond the realm of biosciences and liberal humanism. It is therefore important to map out the assemblages generating problematisations that in turn create the conditions of possibility for particular practices. The third implication is with respect to the broader material turn in critical international relations and security studies. While the position of non-human elements is essential to understanding the relations of power underpinning social assemblages, taking their capacities and capabilities into account is not sufficient. As the preceding analysis has shown, it is also important to gauge how understandings of these and their perceived significance are also conditioned by, and mediated through, culture. The final implication is a gesture towards the continuing relevance of colonial and imperial imaginations to contemporary violence in general and those acts generated through the targeted killing assemblage in particular. This is the spectre haunting the targeted killing assemblage. The people and places most likely to be subject to the kinetic force of the targeted killing assemblage are those people and places that were also subject to some of the worst excesses of colonial and imperial forms of governance historically. This is not a coincidence. The dehumanising properties of capitalism, racism, and Orientalism are well understood.[3] But the replication of these logics, the return to forms of aerial imperialism nearly a century after they were first pioneered, speaks to a tragedy at the heart of liberalism itself: the inability to shed these colonial and imperial pretensions.

The future of the targeted killing assemblage?

While targeted killing and drone warfare continue to draw much scholarly attention, there is still a need to further examine the operations of the kill chain central to the targeted killing assemblage beyond the initial mapping begun here. In addition to ontological questions over what the kill chain might be and its phenomenology, there still remain key sites in the chain that remain under-explored, in part because of the difficulty of accessing the spaces themselves and/or observing those who occupy them. For example, what are the objects, micro-practices, and sensations within ground control station of the drone, in the intelligence hubs that process camera footage, the legal and executive chambers in which approval is granted for personality strikes, and on the ground that link this chain together across time and space? It will also be important to determine if and how command and control structures transform in relation to new RPA capacities like increased autonomy or more pervasive sensor technologies. Similarly, outside of journalistic accounts, very little academic work has examined the assemblage of violence that produces kill/capture missions. While many of the conditions of possibility for such missions fall within the targeted killing assemblage outlined in this book, one suspects that the phenomenological experience would produce some unique elements in comparison to those acts of violence committed through the materiality of the drone.

With drones, their role in targeted killing by liberal regimes is likely to diminish as their use proliferates in other regimes such as Iran or Russia, and/or to non-state actors like Islamic State who will begin to use them for their own strategic objectives. As alluded to with the chemical weapons example above, the previous century has shown that liberal powers have a propensity to create norms against the use of particular forms of violence, or their delivery systems, when they, their clients, or their interests become vulnerable to them. Moreover, key aspects of its legitimating discourse, such as precision and prudence, that help to stabilise the targeted killing assemblage in liberal regimes, will become more difficult when illiberal regimes (or actors) mobilise the same forms of violence. In part, this is because the use by illiberal regimes will undermine notions of civility central to liberal performatives of identity in the targeted killing assemblage. The future of the drone is thus likely to reside in surveillance and less than lethal policing functions central to global pacification efforts that are territorial and extra-territorial in their scope. These developments will not only require continuing investigation into the cultural politics of targeted killing, but also careful analysis of how emerging assemblages of political violence, like cyberwar, are being developed and practiced by liberal regimes on their internal and external others.

Notes

1 It is worth noting that these states are often less forthcoming in providing legal defences for specific cases, generally preferring to keep discussions at an abstract level of hypothetical scenarios.
2 For example, see Said (1978); Gong (1984); and Tuastad (2003).
3 For example, see Nayak (2006); Marx (1972: 93–101); and Feagin (2013).

Bibliography

Berger, J. (2008) *Ways of Seeing*, London, Penguin.

Bourke, J. (1999) *An Intimate History of Killing: Face to Face Killing in Twentieth-Century Warfare*, London, Granta Books.

Derrida, J. (1989/1990) 'Force of Law: The "Mystical Foundation of Authority"', *Cardozo Law Review*, 11(919), 920–1045.

Feagin, J. (2013) *Systemic Racism: A Theory of Oppression*, New York, Routledge.

Gong, G. W. (1984) *The Standard of Civilization in International Society*, Oxford, Oxford University Press.

Gregory, D. (2011) 'The Everywhere War', *The Geographical Journal*, 177(3), 238–250.

Kienscherf, M. (2011) 'A Programme of Global Pacification: US Counterinsurgency Doctrine and the Biopolitics of Human (In)security', *Security Dialogue*, 42(6), 517–535.

Marx, K. (1972) 'Economic and Philosophic Manuscripts of 1844', in R. C. Tucker (ed.), *The Marx-Engels Reader*, New York, WW Norton & Company, 66–125.

Nayak, M. (2006) 'Orientalism and "Saving" US State Identity after 9/11', *International Feminist Journal of Politics*, 8(1), 42–61.

Pasha, M. K. (2005) 'Islam, "Soft" Orientalism and Hegemony: A Gramscian Rereading', *Critical Review of International Social and Political Philosophy*, 8(4), 543–558.

Price, R. (1997) *The Chemical Weapons Taboo*, Ithaca, Cornell University Press.

Said, E. (1978) *Orientalism*, London, Penguin.

Tuastad, D. (2003) 'Neo-Orientalism and the New Barbarism Thesis: Aspects of Symbolic Violence in the Middle East Conflict(s)', *Third World Quarterly*, 24(4), 591–599.

United States Air Force (2007) *Irregular Warfare: Air Force Doctrine Document 2–3*, United States Air Force.

United States Department of Defense (2009) *Unmanned Systems Integrated Roadmap: FY 2009–2034*. Washington, DC: Department of Defense.

Index

Taylor & Francis eBooks

Helping you to choose the right eBooks for your Library

Add Routledge titles to your library's digital collection today. Taylor and Francis ebooks contains over 50,000 titles in the Humanities, Social Sciences, Behavioural Sciences, Built Environment and Law.

Choose from a range of subject packages or create your own!

Benefits for you

» Free MARC records
» COUNTER-compliant usage statistics
» Flexible purchase and pricing options
» All titles DRM-free.

Benefits for your user

» Off-site, anytime access via Athens or referring URL
» Print or copy pages or chapters
» Full content search
» Bookmark, highlight and annotate text
» Access to thousands of pages of quality research at the click of a button.

 REQUEST YOUR **FREE** INSTITUTIONAL TRIAL TODAY | **Free Trials Available** We offer free trials to qualifying academic, corporate and government customers.

eCollections – Choose from over 30 subject eCollections, including:

Archaeology	Language Learning
Architecture	Law
Asian Studies	Literature
Business & Management	Media & Communication
Classical Studies	Middle East Studies
Construction	Music
Creative & Media Arts	Philosophy
Criminology & Criminal Justice	Planning
Economics	Politics
Education	Psychology & Mental Health
Energy	Religion
Engineering	Security
English Language & Linguistics	Social Work
Environment & Sustainability	Sociology
Geography	Sport
Health Studies	Theatre & Performance
History	Tourism, Hospitality & Events

For more information, pricing enquiries or to order a free trial, please contact your local sales team:
www.tandfebooks.com/page/sales